CERAMICS

170101

Industrial Processing and Testing SECOND EDITION

J. T. JONES and M. F. BERARD

IOWA STATE UNIVERSITY PRESS / AMES

JOHN T. JONES is vice-president for research and development of Lenox China, Pomona, N.J., and was formerly associate professor of Ceramic Engineering at Iowa State University, Ames. He holds the B.S. and Ph.D. degrees from the University of Utah. Other employment has been with Coors Porcelain Company, Golden, Colo.; Interpace Corporation, Glendale, Calif.; and Pfaltzgraff Company, York, Pa. He is a member of the American Society for Testing and Materials, International Precious Metals Institute, American Ceramic Society, and Ceramic Educational Council.

MICHAEL F. BERARD is professor and chair of the Division of Engineering Fundamentals and Multidisciplinary Design and professor of Materials Science and Engineering at Iowa State University. He holds the B.S., M.S., and Ph.D. degrees in Ceramic Engineering from Iowa State University. He was previously employed as a materials engineer with Aerojet-General Corporation in Sacramento, Calif., and a ceramic materials specialist on the research staff of the Ames Laboratory of the U.S. Department of Energy. A member of the American Ceramic Society and the National Institute of Ceramic Engineers, he has published extensively in the areas of ceramic processing, point defects, diffusion, phase equilibria, and electrical properties of ceramic materials.

© 1993, 1972 Iowa State University Press, Ames, Iowa 50010
All rights reserved

Authorization to photocopy items for internal or personal use, or the internal or personal use of specific clients, is granted by Iowa State University Press, provided that the base fee of $.10 per copy is paid directly to the Copyright Clearance Center, 27 Congress Street, Salem, MA 01970. For those organizations that have been granted a photocopy license by CCC, a separate system of payments has been arranged. The fee code for users of the Transactional Reporting Service is 0-8138-0291-1/93 $.10.

♾ Printed on acid-free paper in the United States of America

First edition, 1972. *Second printing, 1973. Third printing, 1978. Fourth printing, 1981.*
Fifth printing, 1986. Sixth printing, 1988. Seventh printing, 1989. Eighth printing, 1991.

Second edition, 1993

| Library of Congress Cataloging-in-Publication Data |
Jones, J. T. (John Taylor)
    Ceramics: industrial processing and testing/J. T. Jones and M. F. Berard. — 2nd ed.
      p.   cm.
    Includes index.
    ISBN 0-8138-0291-1
    1. Ceramics. I. Berard, M. F. (Michael F.) II. Title.
TP807.J63   1993
666 — dc20   93-7316

The following are trademarks used within the text:

Ram process is a trademark of Ram, Inc., Springfield, Ohio
LANXIDE and DIMOX are trademarks of Lanxide Corporation, Newark, Delaware
SediGraph is a trademark of Micrometrics, Inc.
MICROTRAC is a trademark of Leeds and Northrup Company
AUTOSORB is a trademark of Quantachrome Corporation
Ro-tap is a trademark of W. S. Tyler, Inc.

Dedicated

to the Memory

of

Adolph Coors III

1915–1960

ONTENTS

PREFACE

Ceramics: Industrial Processing and Testing has been used extensively in ceramic education and in the industry since it was first published in 1972. Many advances have been made in the ceramic industry since that time and for that reason educators (and our publisher) have encouraged us to complete this second edition. As with the first edition, we hope that *Ceramics: Industrial Processing and Testing* will continue to play its role as an introduction for those pursuing careers in ceramic engineering, materials science and engineering, and ceramic engineering technology.

Both students and working ceramic technologists have found the previous book a convenient guide to testing methods and processes. For that reason, the book has been carried from the classroom to the workplace. Readers, some not connected with the ceramic industry, have used the book to become familiar with industrial ceramics.

The book is again divided into two parts: Part 1 is concerned with the scope of the ceramic industry and the processing of raw materials from mining through the manufacturing of ceramic products. A new chapter has been added on selected applications for advanced ceramics. Part 2 is concerned with testing ceramic products and batch calculations and includes descriptions of property measurements most often made on ceramics during production processing. The specialized tools used for making these measurements are described, and precautions in their use are emphasized. Whenever possible, Standard Methods of Test of the American Society for Testing and Materials (ASTM) are referred to as examples of specific procedures. Part 2 should continue to be useful to students in their laboratory work and be of particular interest to technicians and supervisory personnel in the ceramic industry.

As with the first edition, many corporations have helped to provide illustrations to give the reader a visual impression of the industry and test equipment. Barbara Brease, Mickey Herpen, and Fortune Scaraglino were tremendous helpmates in this project. Fortune copied the original text to the word processor so that suitable material could be transferred to the new text. Barbara and then Mickey prepared the text for publication. We thank Randy Geary for several important contributions to this edition.

1

The ceramic industry

THE WORD *CERAMIC* is often used to describe dishes, pottery, and figurines. But wall tile, building brick, high-voltage insulators, and glass products are also ceramics. Further, as engineers know, ceramics are widely used in a host of other specialized applications such as automobiles (not just the spark plug insulator), electric motors, military armor, sensors, nuclear reactors, space vehicles, electronic modules, computers, pumps, valves, metal-processing furnaces and ladles, optical components, lasers, and protective coatings.

Recently, the Nobel Prize was given to two Swiss ceramic scientists for their discovery of ceramic superconductors that function at liquid-nitrogen temperatures. This advancement received heavy media coverage, and many high school and college students have since made such ceramic superconductors in school laboratories. This single event has brought the word *ceramic* to the public eye.

What is a ceramic? Ceramics comprise one of the large classes into which all useful solid materials can be divided, i.e., metals, organics, and ceramics. Perhaps a more complete description of ceramics is that they are products made from natural or synthetic minerals. Most of the important ceramics consist of complex oxides and silicates, although a number of useful carbide, nitride, and boride ceramics are also produced.

Mica is a common mineral that is utilized as electrical insulation—an example of a naturally occurring ceramic. The great majority of ceramics, however, are synthetic materials produced by the careful blending of prepared materials followed by heat treatment to produce new mineral forms. The heat treatment or firing operation can be likened to the natural production of minerals deep within the earth's crust under heat and pressure, except that ceramics produced by firing are the result of careful processing control, in contrast to minerals produced in nature that are the result of chance conditions.

Blending and firing raw materials allows the ceramic engineer to produce useful products including bricks and cements, plasters, abrasives, heat-resistant materials, tableware, glass products of all kinds, and a host of materials having unique electrical, optical, and magnetic properties for use in products that affect almost every aspect of our lives.

Ceramics are high-strength materials by nature, but possess brittle characteristics. They are commonly used in structural applications such as construction materials. The strong chemical bonds occurring in most ceramics result in very stable properties. This strong bonding generally leads to great hardness and high melting temperatures. This stability also helps protect the materials from attack by strong acids and high-temperature corrosive gases and liquid metals.

Ceramics can be produced with densities as high as lead or so low that they will float on water. Some ceramics store electrical charges (for example, capacitors); others are magnetic or are electrical insulators having truly phenomenal resistance to voltage breakdown; still others are electrical conductors. Some newly developed ceramics are true super-conductors which are potentially usable at low, but practical, operating temperatures. The properties given as examples and others have created a most important place for ceramics in modern technology.

Extensive methods are available to characterize the properties of ceramics from the raw materials to the final product. The characterization methods include some developed specifically for ceramics but also involve test methods used in many other fields. Characterization is necessary to assure that the processing of the starting materials is being done correctly and that the final products meet specified property requirements.

The world would be a different place without ceramics. Such critical areas of technology as communications, construction, transportation, power generation and transmission, sanitation, space exploration, and medicine owe their development in part to ceramic technology.

1 Scope of the

CERAMIC MANUFACTURERS in the United States annually market about $40 billion worth of useful products. Only a small fraction of these products are used by the private consumer, the major portion going into key applications in other branches of American industry. The usefulness of ceramics results from the wide range of properties that they possess. The careful processing of starting materials is the most important part of producing final products with optimum and consistent properties. The uniquely beneficial properties of ceramics allow their use in applications where other materials would fail; alternatively, certain detrimental properties limit the scope of their use in some critical applications. The challenge for ceramists today is to increase the window of usefulness for ceramics by improving their properties, thus allowing this class of materials to be used in new applications.

The goal of the producer of ceramic products is always to tailor the properties of the finished product to a specific end use. The secret of this tailoring is careful control of key steps in the production process. For example, glass can be made colorless and transparent, or highly colored and absorbent; it can be made very dense or very lightweight; it can be foamed, spun, blown, pressed, or rolled to shape; it can be made electrically conducting or insulating; it can even be transformed after forming into a crystalline mass having entirely different physical properties from the parent glass.

ceramic industry

The diversity of ceramics makes a comprehensive discussion of the scope of ceramic manufacturing somewhat difficult. For simplification, the ceramic industries can be grouped into several divisions according to the type of products they manufacture. This compartmentalization is not totally artificial, since similar products share analogous production methods and problems. The older, more traditional divisions of the industry are those that produce refractories, glass, whitewares, structural clay products, abrasives, porcelain enamel, and cements. Since World War II the great advances in solid-state science fostered by research in electronics and nuclear physics have resulted in thousands of new and sophisticated ceramic products. These products are categorized as technical ceramics, and they are granted a separate division in the survey in this chapter. In recent years great strides in the science of ceramic processing have resulted in improvements in properties of many types of products, but these developments have had an especially great impact on technical ceramics.

Refractories

Refractories are ceramic products manufactured for the express purpose of withstanding the high-temperature corrosive conditions encountered in industrial furnaces and process vessels. As a class of materials, refractories must have high melting temperatures, good hot strength, resistance to chemical attack, resistance to abrasion, and tailored thermal properties. These materials are fabricated by the refractories manufacturers into furnace linings, crucibles, molten metal ladles, and a host of other special high-temperature products.

The most important users of refractories are the metallurgical industries, the steel industry being the biggest consumer. Although the U.S. steel industry has been challenged in recent years by foreign competition, it is still an enormous industry in the United States and is the backbone of the manufacturing and heavy construction segments of the American economy. Operations in steel making include coking of coal in coke ovens, smelting of iron ore in blast furnaces, and refining of impure molten metal to steel in basic oxygen furnaces; all of this is done in refractory-lined furnaces. The refined molten steel is poured into ladles, which then teem it into ingot molds or, more commonly, into tundishes for feeding continuous casters, and the resulting slabs, ingots, and billets are heat treated between the successive rolling operations needed to produce finished steel products; each of these operations requires many different kinds of refractory products in order to be successful. Ceramic manufacturers also use refractory products of various types in their firing operations. Some are used for supporting or protecting products during firing to high temperatures and others for construction and thermal insulation of the kilns used to fire the ware. Many of the vessels used in petroleum refining are lined with refractories. A flow sheet for manufacturing refractories and a number of examples of refractories and their applications are shown in Figure 1.1.

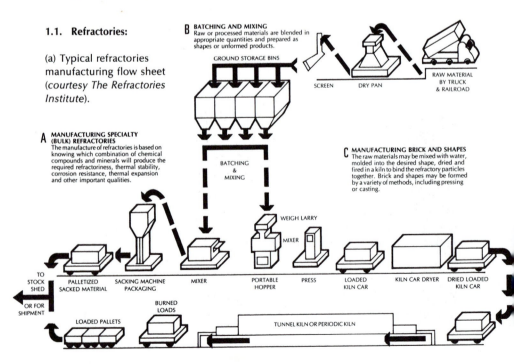

1.1. Refractories:

(a) Typical refractories manufacturing flow sheet (*courtesy The Refractories Institute*).

B BATCHING AND MIXING
Raw or processed materials are blended in appropriate quantities and prepared as shapes or unformed products.

GROUND STORAGE BINS

SCREEN   DRY PAN

RAW MATERIAL BY TRUCK & RAILROAD

A MANUFACTURING SPECIALTY (BULK) REFRACTORIES
The manufacture of refractories is based on knowing which combination of chemical compounds and minerals will produce the required refractoriness, thermal stability, corrosion resistance, thermal expansion and other important qualities.

BATCHING & MIXING

C MANUFACTURING BRICK AND SHAPES
The raw materials may be mixed with water, molded into the desired shape, dried and fired in a kiln to bind the refractory particles together. Brick and shapes may be formed by a variety of methods, including pressing or casting.

WEIGH LARRY

MIXER

TO STOCK SHED

OR FOR SHIPMENT

PALLETIZED SACKED MATERIAL

SACKING MACHINE PACKAGING

MIXER

PORTABLE HOPPER

PRESS

LOADED KILN CAR

KILN CAR DRYER

DRIED LOADED KILN CAR

BURNED LOADS

LOADED PALLETS

TUNNEL KILN OR PERIODIC KILN

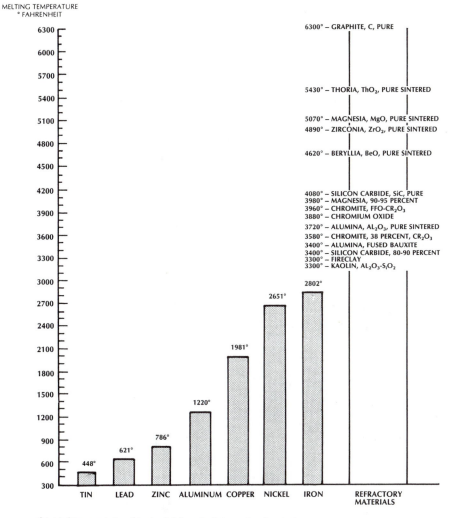

MELTING TEMPERATURE
° FAHRENHEIT

6300° — GRAPHITE, C, PURE

5430° — THORIA, ThO$_2$, PURE SINTERED

5070° — MAGNESIA, MgO, PURE SINTERED
4890° — ZIRCONIA, ZrO$_2$, PURE SINTERED

4620° — BERYLLIA, BeO, PURE SINTERED

4080° — SILICON CARBIDE, SiC, PURE
3980° — MAGNESIA, 90-95 PERCENT
3960° — CHROMITE, FFO-CR$_2$O$_3$
3880° — CHROMIUM OXIDE
3720° — ALUMINA, AL$_2$O$_3$, PURE SINTERED
3580° — CHROMITE, 38 PERCENT, CR$_2$O$_3$
3400° — ALUMINA, FUSED BAUXITE
3400° — SILICON CARBIDE, 80-90 PERCENT
3300° — FIRECLAY
3300° — KAOLIN, AL$_2$O$_3$-S$_1$O$_2$

Material	Temperature
TIN	448°
LEAD	621°
ZINC	786°
ALUMINUM	1220°
COPPER	1981°
NICKEL	2651°
IRON	2802°
REFRACTORY MATERIALS	

(b) Melting points of industrial materials and refractories (*courtesy The Refractories Institute*).

(c) One of hundreds of standard refractory shapes (*courtesy The Carborundum Company*).

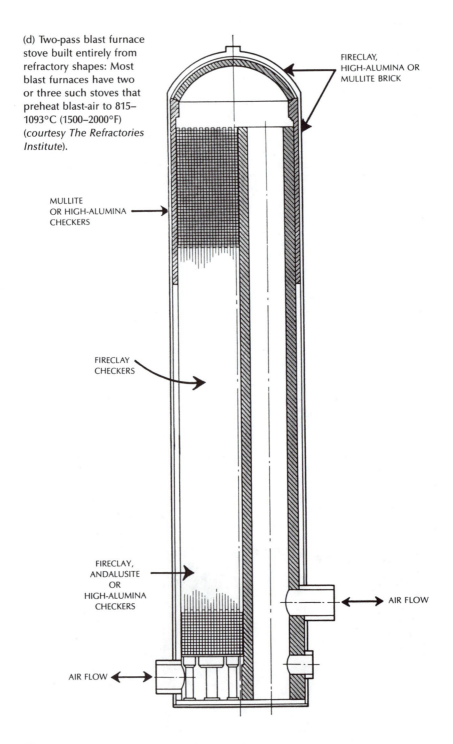

(d) Two-pass blast furnace stove built entirely from refractory shapes: Most blast furnaces have two or three such stoves that preheat blast-air to 815–1093°C (1500–2000°F) (*courtesy The Refractories Institute*).

FIRECLAY, HIGH-ALUMINA OR MULLITE BRICK

MULLITE OR HIGH-ALUMINA CHECKERS

FIRECLAY CHECKERS

FIRECLAY, ANDALUSITE OR HIGH-ALUMINA CHECKERS

AIR FLOW

AIR FLOW

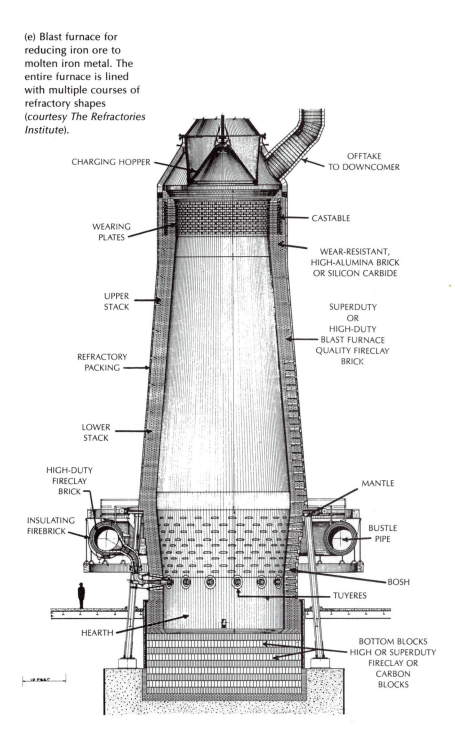

(e) Blast furnace for reducing iron ore to molten iron metal. The entire furnace is lined with multiple courses of refractory shapes (*courtesy The Refractories Institute*).

CHARGING HOPPER

OFFTAKE TO DOWNCOMER

WEARING PLATES

CASTABLE

WEAR-RESISTANT, HIGH-ALUMINA BRICK OR SILICON CARBIDE

UPPER STACK

SUPERDUTY OR HIGH-DUTY BLAST FURNACE QUALITY FIRECLAY BRICK

REFRACTORY PACKING

LOWER STACK

HIGH-DUTY FIRECLAY BRICK

MANTLE

INSULATING FIREBRICK

BUSTLE PIPE

BOSH

TUYERES

HEARTH

BOTTOM BLOCKS HIGH OR SUPERDUTY FIRECLAY OR CARBON BLOCKS

10 FEET

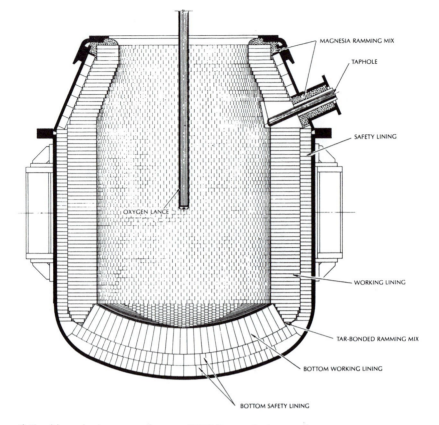

MAGNESIA RAMMING MIX

TAPHOLE

SAFETY LINING

OXYGEN LANCE

WORKING LINING

TAR-BONDED RAMMING MIX

BOTTOM WORKING LINING

BOTTOM SAFETY LINING

(f) Top-blown basic oxygen furnace (BOF) for producing steel:
The lance injects oxygen into the molten metal to burn carbon
and silicon from the steel. Other designs for this furnace
involve oxygen injection through the bottom of the vessel
(*courtesy The Refractories Institute*).

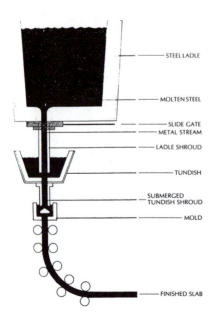

STEEL LADLE

MOLTEN STEEL

SLIDE GATE
METAL STREAM

LADLE SHROUD

TUNDISH

SUBMERGED
TUNDISH SHROUD

MOLD

FINISHED SLAB

(g) Continuous casting of
steel: This process
replaces the casting of
individual ingots.
Refractories (shown in
white and crosshatch)
make the process
possible (*courtesy The
Refractories Institute*).

Refractories are manufactured from a variety of ceramic raw materials including high-alumina fire clays, bauxites, chrome ore, and relatively pure oxides such as silica, alumina, magnesia, and zirconia. Refractory products are also manufactured from graphite, carbides, and silicides. These nonoxide materials are used for specialized applications where their special properties justify their higher cost.

Glass

The tendency of certain ceramic materials not to return to solid crystalline form after being melted and cooled is the basis for the manufacturing of glass. These noncrystalline ceramics behave as if they are very highly viscous liquids, essentially rigid at room temperature but gradually softening and beginning to flow as the temperature is raised.

Most glasses are naturally transparent to light, a property that accounts for many of the uses of glass, such as windows, bottles and other containers, light bulbs, lenses, and optical fibers. Some glasses are opaque to the visible spectrum but are transparent to other wavelengths outside the visible range; such materials can be used for filter and shielding applications.

Although the basic ingredient in most common glasses is silica sand, the chemical composition can be varied over a very wide range by addition of other minerals to the melt to produce a remarkable variety of glasses with tailored properties. Doping the basic glass with oxides of the transition elements produces colored glasses of every imaginable hue. Other additives can render a glass light sensitive, transform it into a laser material, or even make it water soluble.

The viscous liquid nature of hot glass allows it to be formed into products by methods that will not work for other ceramic materials. Glasses can be drawn or rolled into tubes and sheets, blown into hollow containers, pressed into shallow shapes, or spun into fibers. Finally, certain glasses of particular compositions can be heat treated after forming to change them into a crystalline form, producing *glass-ceramics,* which are used in such products as baking dishes, coffeepots, and range tops and also for many technical applications. Figure 1.2 shows a few industrial flat glass products and operations.

1.2. Glass manufacturing (*courtesy of PPG Industries, Inc.*):

(a) A quality control technician visually inspects a continuous ribbon of float process glass as it cools on the way to the cutting station.

(b) Unloaders on a vacuum coating line inspect architectural glass panels onto which metal and metal oxides are vacuum-deposited to give a reflective appearance and enhance energy performance.

(c) Silk-screening equipment is used to apply conductive silver defogger lines to automotive rear windows.

13

Whitewares

Whitewares are the types of products that most people associate
with the term "ceramic." Whitewares as a class of products include electrical and chemical technical porcelains and chinawares, and such commonly used products as fine dinnerware, wall tile, pottery, and dental
porcelains. Figure 1.3 shows some examples of whitewares products and
operations.

Whitewares are usually manufactured from a mixture of minerals
consisting of special clays, fluxing minerals (often feldspars), and finely
ground quartz (SiO_2). This mineral mixture is shaped and then vitrified
(partially melted) at high temperatures to produce a dense, hard material. The fired color is often white (depending on the amounts of impurities in the raw materials) and hence the origin of the general term "whitewares." The whiteness and translucency of so-called bone china is
imparted by adding bone ash to the starting materials. The translucency
is enhanced because the crystalline and glass phases in the fired product
have similar indices of refraction, and this lowers the scattering of light
passing through the china.

1.3. Whitewares products and processes:

(a) A place setting of traditional dinnerware (*courtesy Syracuse
China Corp.*).

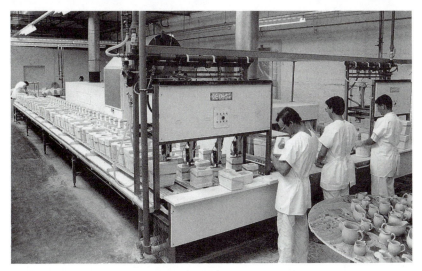

(b) Fabrication of vitreous hotel china hollow ware shapes by slip casting. Ceramic slurry (slip) is first dispensed into plaster molds (center foreground). The part, also called the cast, forms in the mold as it moves down the line to the slip dumping station (far left background). After the slip is dumped, the cast is allowed to set and then the mold with the cast is returned through the drier. The cast is then removed (right center) for finishing and final drying. The time from slip dispensing to removal of the cast from the mold is less than an hour (*courtesy Central Ceramic Services, Ltd.*).

(c) Automatic dry isostatic pressing and finishing of flatware. This horizontal press arrangement forms two or three pieces at once at a rate of 600 pieces per hour. Ware moves from the press (right) to the finishing machine (left rear) and then to the stacking machine, which can sort and stack different shapes separately (*courtesy Dorst-Maschinen and Anlagenbau*).

An important step in the manufacturing of most whiteware products is the application of a glassy coating, called a glaze, to the product. The purposes of the glaze are to seal the surface of the ceramic to prevent absorption of liquids and to increase the aesthetic appeal. The glaze may be colored or colorless, transparent or opaque. Decorations are usually applied over the glaze for household china but under a transparent glaze for institutional ware.

Great advances have taken place in the manufacturing of whitewares between the time when it was an ancient art, steeped in tradition and secret practices, and the present day, when it is a modern technology based on scientific principles and engineering. Important steps in this process of change have been the introduction of automatic processing to replace many of the hand operations of the past, and the development of new body compositions based on high-purity (and sometimes high-cost) starting materials.

Structural clay products

The manufacture of bricks, construction tile, floor tile, vitrified sewer pipe, flue tile, and drain tile is the province of the structural clay products industry. Durability and strength at low cost are key to the success of ceramic structural products. Figure 1.4 shows part of a floor tile manufacturing line and an example of the installed product.

1.4. Ceramic floor tiles (*courtesy American Olean Tile*):

(a) Fast-fire automated floor tile line. The tiles are dry-pressed
from spray-dried powder, then conveyed through this computer-
controlled one-high roller hearth kiln.

(b) Floor tile is used extensively in major construction projects and renovations such as this former railway station, which has been converted into a shopping mall. Glazes allow color variations from the natural fired color of the body.

Structural clay products are usually manufactured from local natural mineral deposits called shales, which are mined by open-pit methods. *Shales* are mineral mixtures rich in clays but having many impurities, including iron-bearing minerals that give them their characteristic brick red color when fired. Most structural clay products are formed by extrusion except for floor tiles, which are usually dry pressed. The successful firing of these highly impure shales often provides the producer with a challenge in temperature control and uniformity. Competition for the construction dollar long ago led this industry to undertake programs resulting in highly automated plants, with many utilizing fast firing cycles in "one-high" tunnel kilns.

Abrasives

Most ceramic materials are hard, but some are extremely hard. The abrasives industry synthesizes ceramic materials of extreme hardness and fabricates them into cutting tools, grinding wheels, and special polishing media (Fig. 1.5). Particularly important ceramic abrasive materials include alumina (also called emery or corundum), silicon carbide, natural

1.5. Special grinding products of the abrasives industry (*courtesy Norton Co.*).

and synthetic diamond, cubic boron nitride, and tungsten carbide. The manufacture of abrasives generally starts by fusing the material into a mass that can then be crushed and sized for particular abrasive applications. Since the fusing operation is normally performed in electric furnaces, abrasives plants are usually located near sources of inexpensive electric power such as dams or, in North America, near waterfalls such as Niagara Falls.

Because most of the important abrasive materials also have very high melting temperatures, abrasives manufacturers also serve as a source of fused ceramic grains for the manufacture of refractories.

Porcelain enamel

Porcelain enamels are glassy coatings fused onto metals to provide corrosion protection and decoration. This industry can be divided into two parts: the producers and the appliers of porcelain enamels. Enamel producers formulate the special glasses that are the major constituent in enamels. After melting, the special glass formulations are quenched and ground into fine powder called a *frit*. Each frit is especially compounded to meet the color, opacity, chemical resistance, and processing requirements of the individual user. The appliers deposit the powdered frits onto specially prepared metal products and fuse them permanently to the surface in a fairly low temperature furnace.

The primary appliers of porcelain enamels are the appliance manufacturers and the fabricators of metal architectural panels and signs (Fig. 1.6).

1.6. Porcelain enameled architectural panels (*courtesy Alcoa*).

Cement and gypsum products

The cement and gypsum industries are concerned with production of the hydraulic-setting cements and plasters used in concrete and other construction materials. Both kinds of materials are produced by careful heat treatment of crushed raw materials to render them reactive with water to form a stonelike mass.

In the cement industry, heat treatment is carried out in immense rotary kilns, which are the largest pieces of moving machinery in the world (Figs. 1.7 and 5.10). The major raw materials for the manufacture of cement are limestone, clays, and iron-bearing minerals. Plaster is produced by gentle heating of gypsum ($CaSO_4 \cdot 2H_2O$) to drive off part of the natural chemical water. When liquid water is added to the plaster, it rehydrates to form massive synthetic gypsum.

1.7. Cement:

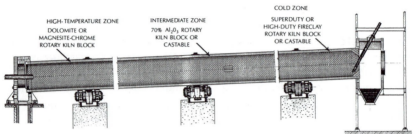

(a) Cement is produced by blending and milling limestone and shale, wet or dry, and calcining the material in a rotary kiln such as the one shown here schematically. Raw materials enter at the upper right while burners fire into the bottom left. As the kiln slowly rotates, the raw materials tumble down through hotter and hotter zones and finally exit at the lower left into a cooling pit (*courtesy The Refractories Institute*).

(b) After cooling, the calcined cement material is finely ground and bagged for shipment (*courtesy Marquette Cement Manufacturing Co.*).

Technical and advanced ceramics

Research in ceramics after World War II made possible the development of a wide variety of nontraditional ceramic products based on high-purity materials fabricated and fired by special techniques. These products are tailored to meet the needs of many industries, including transportation, medicine, electronics, nuclear power, optics, and aerospace.

A catalog of these products is not possible here, but a few items will serve to illustrate the scope of technical ceramics. Ceramics may be made into single electronic components or, especially in the case of silicon and gallium arsenide, can be fabricated into complex multicomponent large-scale integrated circuits. The ferrite family of ceramics has many magnetic applications such as electronic filters and permanent magnets. Ceramic nuclear fuels have replaced most other materials in power reactors. Very high purity oxides are useful as special refractories and electrical insulators. Technical ceramics are important components in rocket nozzle and missile radome applications. Lightweight and high-impact-resistance ceramics have led to the development of ceramic armor for aircraft and military personnel. Figure 1.8 illustrates some technical ceramic products.

1.8. Technical Ceramics:

(a) Alumina ceramics in a variety of complex shapes for operations over a wide range of aggressive conditions including heavy loads, abrasion, and exposure to hot gases, acids, and molten metals and slags (*courtesy The Carborundum Co. and Alcoa*).

(b) Ceramic permanent magnets, also called "ferrites" (*courtesy Indiana General Magnet Products*).

Technical ceramics designed to solve specific critical materials problems are sometimes referred to as *advanced ceramics*. Many different processing techniques are used with specialized starting materials to produce advanced ceramics (Fig. 1.9). These products are used in applications requiring exceptional properties and placing high-performance demands on the ceramic.

Chapter 6 is devoted specifically to advanced ceramics. At this point we will simply list a few advanced ceramic applications to give an idea of what is to come. An example of the use of tailored advanced ceramics occurs in new-concept automobile engines, which benefit greatly from the lower densities and higher heat resistance of ceramic materials as compared to steels. An engine with moving parts of lower mass that are capable of operating at higher temperatures is much more energy efficient than a conventional engine. To have an adequate life cycle in these advanced engines, a ceramic must be tailored to have high strength and toughness along with thermal and chemical durability. Other advanced ceramic products include air/fuel ratio sensors, exhaust catalyst supports, piezoelectric pressure-sensitive transducers, thermistors, phototransistors, varistors, hot plugs, and turbo charger rotors for the automotive industry; magnetic components and inductors, superconductors, capacitors, varistors, packaging, multilayer substrates, optical fibers, lasers, photodectors, light-emitting diodes, solid-electrolyte batteries and fuel cells, and electro-optic components for electronic and photonic applications; microcircuitry for semiconductor applications; reticulated ceramics, hollow spheres, and fibers for refractory applications; whisker-reinforced ceramics, metal-bonded ceramics and transformation-toughened ceramics for cutting tool applications; tissue-compatible bioceramic materials for dental and medical uses; and composites for aerospace and special structural applications.

Developments come rapidly in technical ceramics, and much of today's manufacturing technology will be replaced by tomorrow's innovations.

Ceramic research

The success of both traditional and advanced ceramics depends on the sophisticated application of technology. Basic and applied research have led to engineering advancement in manufacturing methods and products. The history of ceramics shows that, long ago in continental Europe, England, and China, definite testing schemes were used to evaluate trial ceramic compositions with records being kept of significant

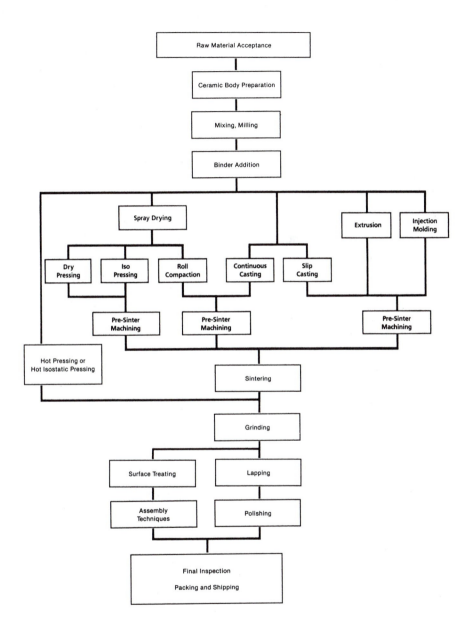

1.9. Various processing routes for producing advanced ceramics (*courtesy Coors Porcelain Co.*).

observations. Because of the importance of ceramics in commerce, the heads of government were intensely interested in promoting ceramic research. This tradition continues today, except that the range of materials and products has greatly expanded, and highly sophisticated analytical instrumentation is now available that the ancients didn't have.

Ceramic research is conducted by private industries, government laboratories, and universities. Universities have traditionally performed the most basic ceramic research, but recent government policies have encouraged the interfacing of universities with industry, and more applied research is also being performed. University ceramic research project listings for about 40 institutions in the United States are published annually in the *Bulletin of the American Ceramic Society*. The research is mainly government supported. In 1992 about 1000 projects were active, with about 800 graduate students involved in the research.

The United States and other Western governments have developed close relationships with industry—mainly because of Japanese competition. Industrial ceramic research is often proprietary, meaning that the results are not widely disseminated as they are for governmental and university research. Much research is wastefully repeated decade after decade because of this secrecy.

In the United States, a key government organization is the National Institute of Standards and Technology (NIST), which performs basic research to obtain an understanding of critical material parameters and develops methods to measure these parameters. The Department of Energy (DOE) has numerous sophisticated materials research facilities that are made available to those who cannot afford them. Such facilities are located at Argonne National Laboratory, Oak Ridge National Laboratory, Sandia National Laboratory, Brookhaven National Laboratory, Lawrence Berkeley Laboratory, and several university DOE laboratories. Laboratories of the U.S. Department of Defense (DOD)—Army, Navy, and Air Force—do extensive research and development on materials of particular interest to the military. The U.S. Bureau of Mines (BM) has traditionally worked to assure the ceramic industry of adequate supplies of raw materials, but it also does basic and applied research in materials processing and development. The National Aeronautics and Space Administration (NASA) supports ceramic research of interest to aerospace applications. The National Science Foundation (NSF) sponsors basic materials research performed at universities.

Ceramic education

At this writing the following universities offered undergraduate and graduate degrees in Ceramic Engineering or Ceramic Science: Alfred University/State University of New York; Clemson University (S.C.); Georgia Institute of Technology; University of Illinois; Iowa State University; University of Missouri, Rolla; Ohio State University; Pennsylvania State University; Rutgers University (N.J.); and University of Washington. A number of other universities have undergraduate programs in Materials Science or Materials Engineering.

Other universities have only graduate programs in Ceramic Science, Ceramic Engineering, Materials Science, or Materials Engineering: Case Western University (Ohio); Clarkson University (N.Y.); University of Florida; Massachusetts Institute of Technology (MIT); North Carolina State University; University of Pittsburgh; Rensselaer Polytechnical Institute (N.Y.); Virginia Polytechnical Institute and State University; Cornell University (N.Y.); University of Southern California; University of California, Berkeley; University of California, Davis; University of California, Los Angeles; Marquette University (Wis.); University of Michigan; Northwestern University (Ill.); Stanford University (Calif.); University of Utah; and Vanderbilt University (Tenn.).

The annual research issue of the *Bulletin of the American Ceramic Society* may be consulted for a listing of other universities doing graduate research in materials.

Hocking Technical University in Ohio has a 2-year program leading to an associate degree in Ceramic Engineering Technology.

Technical societies and publications

Many societies have been organized to promote the advancement of ceramic technology and science. Most technologically advanced countries have one or more technical societies whose members share a common interest in ceramic materials. These organizations publish the results of ceramic research and development and also sponsor conferences for the exchange of information. The publications of these societies are complemented by commercially published trade periodicals that report on developments within the ceramic industries.

A partial list of English-language periodicals devoted solely to ceramic technology follows:

Journal of the American Ceramic Society, Bulletin of the American Ceramic Society, Transactions of the American Ceramic Society,

Ceramic Engineering and Science Proceedings, and *Ceramic Abstracts* (American Ceramic Society, 735 Ceramic Place, Westerville, Ohio 43081)

Transactions of the British Ceramic Society, Proceedings of the British Ceramic Society, and *Journal of the British Ceramic Society* (British Ceramic Society, Shelton House, Stoke Road, Shelton, Stoke-on-Trent, England)

Canadian Ceramic Quarterly (Harold L. Taylor Enterprises, Ltd., 2175 Sheppard Avenue East, Suite 110, Willowdale, Ontario M2J 1W8)

Ceramic Industry (Corcoran Communications Inc., 6200 S.O.M. Center Road, Solon, Ohio 44139)

Glass Technology and *Physics and Chemistry of Glasses* (Society of Glass Technology, Thornton, 20 Hallam Gate Road, Sheffield 10, England)

Interceram (Verlag Schmid GmbH, Freiburg i. Br, Federal Republic of Germany)

The results of ceramic research are also routinely published in government reports and in journals devoted to materials science, solid-state physics, and physical chemistry.

2 Materials used

THE VARIOUS CERAMIC INDUSTRIES in the United States annually process more than 50 million tons of raw materials having an initial value of nearly a billion dollars. These industries use a tremendous variety of inorganic nonmetallic materials ranging in value from a few tenths of a cent per pound for glass sand to several thousand dollars per pound for some of the exotic materials used as additives. Some of these materials are abundant and can be mined, mixed, and used directly, and others are rarer and require extensive chemical processing prior to use; the price varies accordingly. This chapter begins with a description of the characteristics of several dozen of the most common ceramic raw materials making up the bulk of those processed by ceramic industries, followed by a description of a few of the exotic materials. The chapter concludes with a description of mold materials and a note on toxicity.

Characterization of ceramic materials

In the ceramic industry, the properties of the starting materials greatly influence the processing procedures and the nature of the final product. Important factors for starting materials are chemical and mineralogical composition, purity, moisture content, particle size distribution, ultimate particle size in materials of uniform particle size, surface area of the powder, particle density, particle shape, and chemical exchange capacity for clay materials. Methods for determining many of these properties are included in Part 2 of this book.

in ceramic processing

Clays

The word *clay* is used in ceramics to mean several different kinds of material(s). It is often taken to mean one of a particular group of purified true clay minerals, each having a definite composition and a characteristic crystal structure. At other times, it refers to a natural or manufactured mixture of minerals that contains some true clay minerals, which can be made workable by the addition of water.

The true clay minerals occur in great variety in nature and are found in commercially workable deposits of various purities all over the world. True clay minerals are hydrated aluminum silicates that have been formed by alteration and breakdown of parent igneous rocks. The most important source rocks for clay formation are feldspathic rocks or granites (typically mixtures of feldspars, quartz, and mica).

The gradual alteration of parent rocks to form clay minerals has usually taken place over very long time periods within the earth's crust under the chemical action of hot, high-pressure gases and water. Deposits of clays formed by such a hydrothermal process are frequently found still mixed with fragments of unaltered parent rock. These are called residual or primary clay deposits.

Aeons ago, erosion exposed many primary clay deposits, and streams and rivers washed much of the clays out from their original deposits. The action of the moving water tended to partially purify the material by separating out some of the heavier parent rock fragments. The resulting finer particles were washed downstream to settle out in the calm waters of lakes or where rivers emptied into ancient seas. Clays deposited in such a manner are called sedimentary or secondary clays. These deposits still contain some finely divided mineral impurities and often also contain organic material resulting from decaying vegetation entrapped in the clays as they settled out of the water.

The most abundant true clay mineral is kaolinite, which is a hydrated aluminum silicate with the chemical formula $Al_2Si_2O_5(OH)_4$. A clay raw material that consists primarily of kaolinite is called a kaolin. The name *kaolin* comes from the Chinese word *kauling,* which means "high-ridge," the name of a hill near Jauchau Fu where an important ancient deposit was located. Excellent secondary deposits of kaolin are found in the southeastern United States, the most important locations being in the Carolinas, Georgia, and northern Florida. Major primary deposits are found in Cornwall, England, where it is referred to as "china clay." Many countries throughout the world have workable kaolin deposits. Figure 2.1(a) is a photograph of dispersed kaolinite crystals taken with an electron microscope. This photograph reveals that the form of kaolinite crystals is that of thin platelets, roughly hexagonal in shape. Kaolinite occurs in nature in relatively thick beds made up of billions of these tiny kaolinite crystals, which typically measure about 1 millionth of a meter (1 μm), or approximately 40 millionths of an inch, across the plate face by about 0.1 μm in thickness. Several structural variations of the fixed kaolinite formula exist, depending on differences in internal arrangement of the Al, Si, and O atoms in the crystal but they are less commonly used in ceramics. The form called halloysite, shown in Figure 2.1(b), is used in some whiteware formulations. Note that here the platelets have rolled up into tubes. The halloysite material used in whitewares comes mainly from New Zealand.

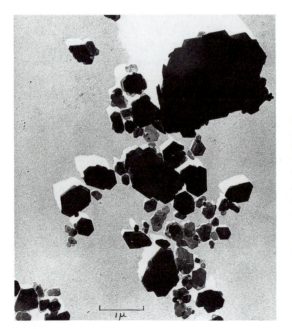

2.1. Kaolin type minerals:

(a) Kaolinite particles are in the form of thin hexagonal platelets as shown in this transmission electron micrograph. The bar (1 μ) indicates a length of one millionth of a meter or about 0.00004 inches (*courtesy Georgia Kaolin Co.*).

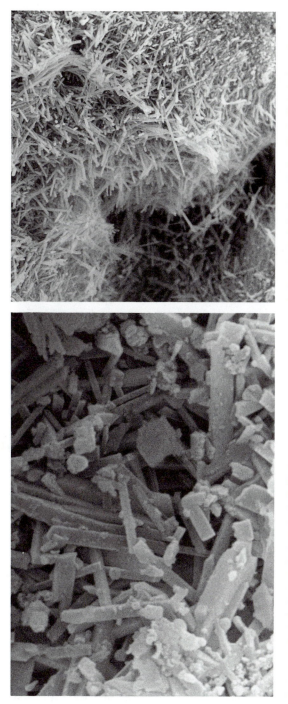

(b) Halloysite contains some water, which is easily removed by gentle calcination. The composition then becomes similar to that of kaolinite. Note that the platey particles have "rolled up" into tubes. Excellent ceramic-grade halloysite is mined in New Zealand. The halloysite shown in this particular scanning electron micrograph is from a deposit at the eastern end of the Mediterranean Sea (*courtesy ECC International*).

(c) Heating of kaolins to high temperatures results in the development of mullite ($3Al_2O_3 \cdot 2SiO_2$), an important mineral in all fired whitewares and fire clay refractories. The mullite shown here in a scanning electron micrograph is from calcined English china clay. The specimen was etched in strong alkali to remove the glass phase so that the long prismatic crystals of mullite can be observed (*courtesy ECC International*).

Additions of water to a kaolin cause it to become *plastic*—that is, the resulting thick paste can be deformed and molded, and it will hold its new shape. This unique property of clays in general is the basis for many of the hand- and machine-forming methods used in the production of clay-containing ceramic products. (See Chapter 4 for forming methods.) The degree to which a particular clay shows plastic behavior is a complex function of particle size, particle shape, and impurity content. Kaolins from different deposits vary enough in particle size and chemical and mineral composition to make substitutions of one for another in batching possible only after careful testing. Such substitutions can make significant differences in the amount of water required for processing as well as changes in the fired product. Usually reformulation of the body rather than simple substitution is required, as described in Chapter 8.

Kaolins fire to a white or nearly white color. This property, combined with their ability to be plasticized, formed, and then fired into a dense, hard mass in the presence of fluxes (substances that help generate liquid and thereby lower the required firing temperature), makes kaolin a prime ingredient in many traditional ceramic products. Mullite, $3Al_2O_3 \cdot 2SiO_2$, develops from the thermal decomposition of kaolin (Fig. 2.1[c]). This mineral is an important constituent in fired clay-based ceramics, as is discussed in Chapter 4.

Particularly fine-grained secondary deposits of kaolinite having especially high plasticity and stickiness when mixed with water are called *ball clays*. The great plasticity of ball clays makes them useful as additions to somewhat less plastic kaolins to improve workability. As found in nature, ball clays are often dark in color due to the presence of finely divided organic material, which probably aids in improving their plasticity. The organic material is lignite, with properties intermediate between peat and bituminous coal. Excessive lignite is often removed during processing. Ball clays fire to give a cream-to-white color. Because of the large amounts of water that must be added to develop their high plasticities, ball clays will exhibit large shrinkages during drying and therefore cannot completely replace kaolins in a ceramic body without causing cracking and warpage of the ware. Ball clay deposits exist in many countries. The chief deposits of ball clays in the United States are found in Kentucky and Tennessee and those in England are in Devon.

Montmorillonite is a true clay mineral with a crystal structure completely different from kaolinite but fairly closely related to the structure of the mineral mica. The general chemical formula for montmorillonite is $Al_2Si_4O_{10}(OH)_2$ but there are always significant amounts of iron, magnesium, and potassium atoms present within the crystals. Pyrophyllite, sometimes used as an ingredient in ceramic wall tile, has the same general chemical formula with almost no impurity atoms. The crystal structure of montmorillonite consists of layers, and large amounts of water can be accommodated between the layers, leading to the possibility of considerable swelling when the dry clay contacts water. Montmorillonite is the major mineral in deposits of very fine-grained, highly weathered, volcanic glass called bentonite. Bentonite is mined in the United States in Wyoming and South Dakota. A mixture of bentonite and water is very sticky and is sometimes added in small amounts to ceramic formulations to improve workability and green (unfired) strength.

Illite is a true clay mineral fairly closely related to mica and is a common constituent in many of the clay or shale deposits mined throughout the United States by the structural clay products industry. *Shales* are naturally occurring compressed mixtures of clay, sand, and other minerals; they can be ground and, because of their clay content, can be made plastic with water to manufacture brick, tile, and sewer pipe. The brown-to-red fired color of these products is a result of iron-bearing minerals in the shale. The common presence of water-soluble sulfate minerals (like gypsum) in these deposits is usually overcome by the addition of barium salts to render the sulfates insoluble and thereby to prevent surface scumming during drying of the ware.

Fire clays are deposits that are predominantly kaolinite but with additional minerals such as diaspore which raise the total alumina (Al_2O_3) content above that of normal kaolinite. These clays are often, although not always, fairly plastic and normally fire to a buff color. The chief use of fire clays is in the production of many important refractory products used by industry to control and contain high temperatures. The most important deposits of fire clays in the United States are found in central Missouri, although numerous other deposits are scattered over the country.

Important characteristics often determined for clays intended to be used in ceramics are particle size and shape, surface area of particles, chemical composition, mineralogical composition, and cation exchange capacity. Typical data from suppliers are shown in Table 2.1.

Talc is a claylike hydrated magnesium silicate mineral with the chemical formula $Mg_3Si_4O_{10}(OH)_2$ which is similar in structure to pyrophyllite. The soft, almost soapy-feeling mineral is used as a major constituent in steatite porcelains, especially for wall tile, art pottery, and low-loss electrical insulators. Talc is also an important constituent in cordierite porcelains, which are known for their very low thermal expansion. Important deposits of talc in the United States occur in the Southern California–Nevada region, the North Carolina–Georgia region, and in Montana (see Table 2.2 for properties).

TABLE 2.1. Representative clay materials—typical chemical analysis and particle size

Material	SiO$_2$	TiO$_2$	Al$_2$O$_3$	Fe$_2$O$_3$	CaO	MgO	K$_2$O	Na$_2$O	LOI[g]	C	% <2μm
English ball (ESVA)[a]	49.8	1.00	32.2	1.00	0.20	0.40	2.40	0.30	12.80	2.3	85
English ball (EWVA)[a]	47.6	0.90	33.1	1.00	0.20	0.30	1.70	0.10	15.10	3.1	80
Georgia kaolin (Pioneer, air floated)[b]	45.7	1.43	38.5	0.44	0.24	0.14	0.14	0.04	13.50	...	55–65
Georgia kaolin (No. 6 tile, air floated)[b]	46.9	1.42	38.2	0.35	0.43	0.58	...	0.40	13.90	...	54–65
Tennessee ball (Volunteer)[c]	58.9	1.78	27.1	1.13	0.04	0.14	0.45	0.18	10.28	...	72.6
Tennessee ball (Jackson)[d]	54.9	1.70	30.0	1.00	0.30	0.40	0.30	0.10	11.30	...	85
Kentucky ball (Old Mine-4)[d]	55.1	1.20	27.9	1.10	0.30	0.40	1.00	0.30	12.60	...	81
China clay (SSP)[e]	47.0	0.03	38.0	0.39	0.10	0.22	0.80	0.15	13.00	...	85
China clay (Grolleg)[e]	48.0	0.02	37.0	0.70	0.06	0.30	1.85	0.10	12.20	...	57
Tennessee ball clay (Bandy Black)[f]	61.0	1.29	24.5	0.99	0.09	0.12	1.69	0.36	9.74	...	44
Georgia kaolin (Velvacast®)[b]	45.4	1.43	38.9	0.34	0.24	0.18	Trace	0.11	13.81	...	38–40
Georgia kaolin (Kaopague®)[b]	45.2	0.66	39.3	0.32	0.21	0.03	0.09	0.03	13.92	...	...

[a] WBB, Ltd.
[b] Georgia Kaolin Co.
[c] Old Hickory Clay Co.
[d] K-T Clay Co.
[e] ECC, Ltd.
[f] Spinks Co.
[g] Loss on ignition.

TABLE 2.2. Feldspathic materials, talc, and bone ash—typical chemical analysis

Material	SiO$_2$	TiO$_2$	Al$_2$O$_3$	Fe$_2$O$_3$	CaO	MgO	K$_2$O	Na$_2$O	P$_2$O$_5$	LOI
K-spar	67.70	...	18.00	0.07	0.14	trace	10.10	3.60	...	0.30
Na-spar	68.13	...	18.70	0.07	1.50	trace	4.60	6.80	...	0.20
Nepheline syenite	60.70	...	23.30	0.07	0.70	0.10	4.60	9.80	...	0.70
Calcined bone ash	1.73	0.01	0.13	0.08	53.00	1.12	0.02	0.60	39.90	2.74
Cornish stone	72.40	0.05	15.30	0.12	1.41	0.16	4.32	3.44	0.47	2.15
Talc (typical)	56.00	...	0.31	0.16	8.19	30.00	...	0.34	...	5.00
Pyrophyllite	81.00	0.01	13.70	0.20	...	...	2.30	0.40	...	2.30
Wollastonite	52.00	...	1.80	0.34	42.10	1.49	...	0.27	...	2.00

Fluxing materials

A *flux* is a material that appreciably lowers the required firing temperature of a ceramic by reacting with other materials present at fairly low temperature to form molten glass. Many inorganic materials exhibit some fluxing tendency, but the most important fluxes for ceramics are those that contain alkalies (Li, Na, K, Rb, Cs), alkaline earths (Ca, Mg, Sr, Ba), boric oxide, the lead oxides, or fluorine. Any mineral that contains appreciable amounts of these substances, usually as oxides or silicates, is therefore a fluxing mineral.

The feldspars (or spars) are a group of minerals that form molten glass at moderate temperatures. All of the spars of importance to ceramics consist of aluminosilicates of sodium, potassium, or calcium. The pure spars are albite ($NaAlSi_3O_8$), orthoclase ($KAlSi_3O_8$) and anorthite ($CaAl_2Si_2O_8$). These minerals often occur together in nature. Specifically, the minerals albite and anorthite can be found in solid solution in any proportion in the so-called plagioclase series of feldspars. On the other hand, very little solid solution occurs between anorthite and orthoclase. The most important feldspars for routine ceramic applications are mixtures of albite and orthoclase. These spars are often specified by their soda to potash ($Na_2O:K_2O$) ratio.

Large deposits of feldspars are found in New England and North Carolina in the United States and also in Canada. Minor deposits of good quality are found in several other states. Since feldspars are mined by blasting from massive rock formations, the rock must be crushed and milled so that the feldspar can be separated from other minerals such as mica, quartz, and rutile by chemical flotation and magnetic separation. In Great Britain, certain partially decomposed granites (rich in feldspars) called *china stones* are used as a flux in whiteware bodies. Local varieties such as Cornish stone (from Cornwall) (Fig. 2.2) and Manx stone (from the Isle of Man) are examples, the former containing some CaF_2 (fluorite or fluorospar), a powerful flux that gives the unfired material a purple cast. Typical feldspar and Cornish stone properties are shown in Table 2.2.

Nepheline syenite is a fluxing material consisting of about 75% mixed soda and potash feldspars and about 25% of the mineral nepheline ($NaAlSi_2O_4$). Major deposits are found in Ontario, Canada. The type of iron-bearing secondary minerals present varies with deposit and this affects the supplier's ability to lower the iron content by magnetic separation after the rock is crushed. Typical properties of nepheline syenite are shown in Table 2.2.

Soda ash is a fluxing mineral of the formula $Na_2CO_3 \cdot xH_2O$, where x can take several values such as 1, 7, or 10. This material is the primary source of the soda (Na_2O) flux used in the production of flat and container glass. Much soda ash is taken from natural desert alkali deposits, but the material is also produced by chemical processing of brines.

Borax is a hydrated sodium borate mineral having the chemical formula $Na_2B_4O_7 \cdot 10H_2O$. Since this mineral thermally decomposes to give two potent fluxes, soda (Na_2O) and boric oxide (B_2O_3), it is a very powerful fluxing additive. The major uses of this mineral are in glasses, glazes, and enamels. Dried lake beds in California often contain quite pure deposits of borax and they serve as the chief sources of the mineral.

A number of lead oxides are used as fluxes in glazes and enamels and also as major constituents in the brilliant lead glasses used to produce so-called glass "crystal" and optical glasses. The important lead oxides used in ceramics are litharge (PbO), red lead (Pb_3O_4), and white lead (also called basic lead carbonate), $2PbCO_3 \cdot Pb(OH)_2$. These synthetic materials are products of the chemical-processing industry and their use is continually under government scrutiny because of their toxicity, especially to children and to women of child-bearing age.

Silica

Silica (SiO_2) in the form of the mineral quartz is the second most utilized ceramic raw material after the clays. It is the chief constituent in most glass batches and is a major ingredient in most glazes and porcelain bodies. Quartz is the most common mineral in the earth's crust, but is often found tightly bound up with other minerals in granites and other igneous rocks. However, large deposits of sandstone of amazing purity occur in Pennsylvania, Illinois, and West Virginia, with numerous other deposits scattered through the Great Lakes states and California. These sandstones consist of a mass of loosely cemented quartz grains and are readily reduced to pure quartz sand by hydraulic mining techniques.

The glass industry uses enormous quantities of silica glass sand resulting from the breakdown of very pure sandstone. The silica used to impart strength and stability to porcelain bodies is a much finer-grained form of quartz called *potter's flint,* which is produced by fine grinding of virtually iron-free quartz sands. Potter's flint is also used as a glass-forming ingredient in glazes and enamels. Nearly pure coarsely granular silica is also molded into refractory shapes that are especially resistant to attack by acidic (silica-rich) slags. Special measures are required to control silica dust in the workplace due to the risk of silicosis from inhalation of the dust over long periods of time.

Refractory oxides

The high melting points and resistance to chemical attack of several pure oxides make them useful as refractories where fire clays do not provide satisfactory service. These oxides are also often used as additives to increase the refractoriness of a ceramic mixture.

Alumina (Al_2O_3) is used extensively in the manufacture of technical ceramics. In essentially pure form, it is an excellent electrical insulator and abrasive, and its high melting point (2050°C, 3722°F) makes it an excellent refractory. Alumina is sometimes used as an additive to replace part of the potter's flint in dinnerware compositions to allow faster firing cycles and to increase strength. It is also used as an ingredient in heat-resistant glasses. Ceramic-grade alumina is a synthetic material produced by the Bayer process involving chemical digestion of natural, impure alumina hydrates (bauxites) followed by crystallization and calcination to remove water. Bauxites are mined in Arkansas, Missouri, Australia, France, the People's Republic of China, and many countries in the Caribbean area. Typical properties of calcined aluminas produced by the Bayer process are shown in Table 2.3.

Fused alumina, formed by melting Bayer alumina or bauxite in an electric-arc furnace, has long been used for refractory and abrasive applications. A bubble form of fused alumina has good thermal insulating properties and is lightweight. *Tabular alumina* is formed by heating pelletized Bayer alumina to 1870°C (3398°F) and has properties approaching those of fused alumina. Typical properties of tabular and fused aluminas are included in Table 2.3.

The so-called basic refractory oxides — magnesia (MgO), chromia or chromium oxide (Cr_2O_3), and lime (CaO) — have long been important for manufacturing furnace-lining materials for the steel industry. With the advent of the basic oxygen process in steel making, the need for these refractories increased tremendously. Magnesia can be produced by hard calcining, or "dead-burning," the mineral magnesite ($MgCO_3$), but the purest material is produced from seawater or deep-well brines by precipitation of dissolved magnesium as $Mg(OH)_2$. In either case, the mineral produced by high-temperature calcination is periclase (MgO), an extremely stable, high-melting-point material. Lime is produced by calcination of natural limestone ($CaCO_3$). (A very pure synthetic form of $CaCO_3$ called *whiting* is a common ingredient in ceramic bodies and glazes.) A mixture of lime and magnesia for use in refractories is produced by calcination of the mineral dolomite ($CaCO_3 \cdot MgCO_3$). Chrome-ore refractories are generally made from complex mineral mixtures containing chromite, which is chromium-iron spinel ($FeCr_2O_4$), but

TABLE 2.3. Typical chemical properties of fused, calcined, and tabular aluminas

Material	Avg. Particle Size μm	SG	Al_2O_3	SiO_2	TiO_2	Fe_2O_3	CaO	MgO	Na_2O
Fused	N/A[b]	3.90	99.5	0.060	0.02	0.04	...	0.007	0.07
Calcine A	5.0	3.80	99.2	0.020	...	0.04	...	...	0.45
Calcine B	8.0	3.85	99.5	0.080	...	0.04	...	...	0.10
Calcine C	3.5	3.85	99.6	0.120	...	0.04	...	...	0.20
Calcine D	1.5	...	99.7	0.050	...	0.04	0.03	...	0.04
Calcine E	0.4	...	99.5	0.025	...	0.01	...	...	0.08
Tabular	N/A[b]	3.96	99.5	0.060	...	0.06	...	...	0.10

[a]SG = specific gravity.
[b]N/A = not applicable.

TABLE 2.4. Typical composition of Virginia kyanite

Materials	SG[a]	SiO_2	TiO_2	Al_2O_3	Fe_2O_3	CaO	MgO	Na_2O/K_2O	LOI
Raw kyanite	3.6	40.9	0.67	57.0	0.72	0.03	0.01	0.42	0.21
Calcined @ 3000°F (1649°C) (mullite)	3.0	49.9	0.67	57.3	0.72	0.03	0.01	0.42	0.00

Source: Kyanite Mining Corporation.
[a]SG = specific gravity.

TABLE 2.5. Typical chemical properties of high-purity aluminas

Material	Nominal Particle Size μm	Purity	Crystalline Form	Surface Area (m²/gm)	SG[a]	Starting Material
A	0.30	99.980	Hexagonal	14	3.94	Alum
B	0.05	99.980	Cubic	82	3.72	Alum
C	1.00	99.980	Hexagonal	3	3.98	Alum
D	...	99.999	Gamma[b]	55	...	$AlCl_3$
E	...	99.999	Alpha (Hex)	<6	...	$AlCl_3$
F	1.00	99.980	Alpha (Hex)	22	...	$AlCl_3$

[a]SG = specific gravity.
[b]Low-temperature hydrated form.

always also containing some Mg, Al, and Si impurities. The refractoriness of chrome ore is often improved by the addition of MgO prior to firing. No deposits of chromite are found in the United States. The use of this mineral has become somewhat curtailed by the recent discovery of the toxicity of certain valence states of dissolved chromium.

Miscellaneous conventional ceramic raw materials

In addition to the materials already described, several hundred other raw materials are used by the various ceramic industries. The tonnages used of these materials are considerably less than for the major materials already mentioned, but their dollar value is certainly not insignificant. Several of these minor materials are sufficiently common to warrant mention here.

Graphite is essentially pure carbon. It is a superrefractory material that will not melt even at the highest industrial temperature. Graphite is generally not wetted by melts and slags and therefore resists chemical attack; however, it must be protected from oxidation above red heat. Graphite can be found in nature, but the greatest quantity is produced synthetically from petroleum coke. Furnace liners, crucibles, casting molds, and electric furnace electrodes are the major ceramic uses of graphite. It is often mixed with fire clay to produce crucibles that are resistant to oxidation.

The sillimanite minerals all have the chemical formula Al_2SiO_5. Different crystalline forms are called kyanite, sillimanite, and andalusite (Table 2.4). These natural minerals, because they are rich in Al_2O_3, are used to increase the refractoriness of silicate ceramic compositions.

Silicon carbide (SiC) is an excellent refractory and abrasive material. It does not occur naturally, but is produced synthetically by reacting silica sand and petroleum coke at very high temperatures in an electric-resistance furnace. When formed into refractory shapes, it may be used either alone or in a mixture with fire clay. Silicon carbide fibers and whiskers can be used to reinforce metals and other ceramics such as alumina, silicon nitride, and mullite. Improvements in toughness, thermal shock resistance, wear resistance, elasticity, and strength can be attained in these composite materials. SiC dust can be a serious contaminant if inadvertently introduced into whiteware operations and must be controlled.

Zirconia (ZrO_2) and zircon ($ZrSiO_4$) are important refractory materials and excellent thermal insulators. Zirconia, which is produced by chemical means from natural zircon, is seldom used in the pure form, but rather is usually stabilized by the addition of lime or other oxides to eliminate destructive crystal structure changes during heating and cooling. Zircon is often used as an opacifier in glazes to increase their covering power, that is, to decrease the thickness of glaze needed to be opaque.

Materials for advanced applications

High-purity aluminas
Alumina of greater purity than Bayer grades can be synthesized from very pure starting materials such as alums (for example, $NH_4Al(SO_4)_2 \cdot 12H_2O$), aluminum chloride, or even aluminum metal. Carefully controlled processing allows the manufacturer to retain the purity of the product. This material is used to produce such products as translucent alumina tubes for sodium vapor lamps. There are special electronic and mechanical applications for high-purity aluminas as well. Transformation toughened aluminas (TTA) can be produced by the addition of 8 to 15 volume percent zirconium oxide. Aluminum titanate, formed when alumina and titania are reacted together at about 1300°C (2372°F), is useful for automotive applications where easy machining of formed parts and high heat resistance are needed. Table 2.5 illustrates typical properties of high-purity alumina powders.

Barium titanate
Barium titanate ($BaTiO_3$) is a synthetic material used in capacitors, sonar equipment, ultrasonic cleaners, flaw detection equipment, and many other applications requiring special dielectric, piezoelectric, or ferroelectric properties. The structure of the material accommodates extensive substitution of other ions, which allows additives to be used to change the properties of the material dramatically. Mixing and sintering — or better, hydrothermal processing — can be used to produce a whole range of titanate powders. Two important materials in the titanate family are PZT (lead zirconate titanate) and PLZT (lead lanthanum zirconate titanate). These compounds have special dielectric and piezoelectric properties that are very useful in producing capacitors and transducers.

Beryllium oxide
Beryllium oxide (BeO), or *beryllia,* extracted by chemical processing of the mineral beryl ($3BeO \cdot Al_2O_3$) is used in applications where its high thermal conductivity, good dielectric properties, and/or special nuclear

properties are needed. This material can be very toxic in the powder form and special precautions are required during manufacturing.

Boron carbide

Boron carbide (B_4C) is a very hard and very refractory synthetic material used for grinding, drilling, and polishing applications. It can be bonded into an abrasive shape with metals (cermets) or other materials, but this usually diminishes its properties. Boron carbide reacts with oxygen and molten transition element metals at high temperatures.

Boron nitride

Boron nitride (BN) is a most promising synthetic material. It has properties somewhat like graphite, including good refractory properties and resistance to wetting by molten metals and slags; however, unlike graphite, it is a good electrical insulator. Its high dielectric strength, high thermal conductivity, low thermal expansion, and good strength retention when heated make it an excellent substrate material for electrical circuitry. A special cubic form of BN that rivals the hardness of diamond can be made by high-pressure techniques.

Diamond

Diamond is produced synthetically now in commercial quantities and finds application in the electronic industry as well as for abrasives. Diamond abrasives are used extensively in the machining of ceramics. Recently, techniques for depositing diamond coatings on other materials have been developed.

Rare earth oxides

Rare earth oxides are produced from ores by complex chemical separation techniques. Dysprosium and gadolinium oxides are used in nuclear and dielectric applications; erbium oxide is used as a glass colorant; europium oxide is used in nuclear control rod applications and is a constituent in the standard TV red phosphor; lanthanum oxide is used in special glasses and as an additive in PLZT compositions; neodymium oxide is used in special glasses and single crystals as the crucial ingredient to impart laser behavior; praseodymium oxide is used as a colorant for glass (green) and for glazes (yellow); samarium oxide is used as a phosphor additive and in luminescent and infrared absorbing glasses. In general, all of the rare earth oxides can be used as dopants in a variety of dielectric materials to change their electrical and optical properties.

Ferric oxide

Ferric oxide (Fe_2O_3) is used in compounding ceramic magnetic materials called *ferrites*. This oxide is blended and processed with other oxides such as BaO, ZnO, and MgO to form permanent (hard) magnets or soft magnets (which have little residual magnetization) according to end use.

Silicon nitride

Silicon nitride (Si_3N_4) is a synthetic material that shows excellent hot strength, corrosion and oxidation resistance at high temperatures, and resistance to attack by nonferrous molten metals. This material is being used in gas turbine engines and piston engine parts and in antifriction applications.

Titanium carbide and boride

Titanium carbide (TiC) is a very hard, refractory material used in cutting tools. Titanium diboride (TiB_2) has similar properties and has a number of applications including ceramic armor, bearing parts, cutting tools, and special refractories.

Materials for ceramic molds

Gypsum is a natural mineral having the formula $CaSO_4 \cdot 2H_2O$. When heated gently, part of the water of hydration is removed to yield plaster. When water is added to powdered plaster, the material gradually "sets" by rehydration back to porous gypsum. Plaster is a useful construction material, but its water absorption ability also makes it an important material for use in ceramic forming molds, as will be explained in detail in Chapter 4.

Various acrylics and other polymeric materials can be formed into porous molds which mimic the water-absorbing action of plaster on ceramic slurries. These will be discussed later as they relate to particular forming processes. Porous metals have also been used, at least experimentally, to form ceramics.

Toxicity of ceramic materials

Toxic raw materials are federally regulated in the workplace by the Occupational Health and Safety Administration (OSHA), in the environment by the Environmental Protection Agency (EPA), and in the marketplace by the Food and Drug Administration (FDA). Lead compounds, silica, beryllium oxide, and many other ceramic materials come under various regulations. Suppliers of raw materials are required to provide a Materials Safety Data Sheet (MSDS) to users. These sheets reveal the hazardous ingredients in a material, physical data related to use, fire and explosive data, and health hazard data including exposure limits, reactivity data, decomposition products, spill or leak handling procedures, protection information, and special precautions. Additional information often provided by governmental agencies concerning specific health risks are also given in the MSDS. Literature and vendor sources for information on raw materials do not usually list their toxic properties. A good source for toxicity information is the latest edition of *Dangerous Properties of Industrial Materials,* N. Irving Sax, Van Nostrand Reinhold, New York. Ceramic raw materials should be treated as if they were chemicals, which, in fact, is what they are. While many common ceramic raw materials have relatively low toxicity ratings, none should be handled until the user has determined the hazards that may be involved.

Raw materials

MANY CERAMIC RAW MATERIALS, especially those of modest purity that are used in large tonnages, are mined from large deposits in the earth's crust. The manufacture of ceramic products really begins with prospecting for these mineral deposits, and often includes such operations as constructing roads and running utilities to the mine site. After mining, the materials must be purified, crushed, sized, and transported to the manufacturing site. Here further crushing and sizing usually take place, followed by blending with other raw materials and conditioning with water or other additives. Only then can a useful product be formed from the raw mix. The first part of this chapter is concerned with the operations performed in traditional ceramic manufacturing from the mining operation up to the forming step. A great variety of processing equipment is involved in these operations.

Ceramics for advanced applications often start not with mined materials, but rather with specially synthesized materials, to assure the added degree of purity and property control needed in the final product. If the material synthesis is not performed properly, further processing will not bring the final product within required specifications. For this reason, advanced ceramic manufacturers often start with basic chemistry to make needed materials. Synthesis processes in reasonably common use are described in the last part of this chapter.

processing

Mining

Ceramic raw materials may be taken from the earth in several different ways (Fig. 3.1). The most important method is strip mining or open-pit mining, and is useful when materials are located close to the surface in large, uniform deposits. The earth covering the deposit (called over-

3.1. Many ceramic raw materials, including clays, are mined by open-pit methods:

(a) After stripping away the overburden, heavy machinery is used to collect and remove the clay (*courtesy H.C. Spinks Co.*).

(b) Kaolin (china clay) recovery by hydraulic mining methods in Cornwall, England (*courtesy ECC International*).

(c) Classification of clay slurries in large settling tanks. Slurries from different mines may be blended to achieve desired chemical, physical, and ceramic properties (*courtesy ECC International*).

(d) When the pit is exhausted, the land is reclaimed for agricultural or recreational uses (*courtesy H.C. Spinks Co.*).

burden) is removed with bulldozers, and the raw material is usually collected with standard earth-moving equipment. In some cases, such as English china clays and loosely cemented silica sandstones, the exposed material can be washed out from the deposit with high-pressure hoses (sometimes called placer or hydraulic mining). In the case of deposits of massive hard rock such as feldspar and nepheline syenite, after the overburden is removed, hard-rock mining techniques are used, including blasting followed by sorting and crushing at the mine site. After open-pit mining, the land must be reclaimed and converted to pasture or other uses. Artificial lakes are sometimes formed and converted to recreational purposes.

When the mineral deposit is located far below the surface, vertical shafts or horizontal tunnels must be driven into the deposit. Material is often removed by drilling and blasting. The tunnels must usually be supported or shored up to prevent cave-ins. Miniature railroad systems can often be used to carry raw materials horizontally to the surface; at other times, the raw materials must be transported to the surface in vertical conveyors or elevators.

To minimize the cost of transportation of raw materials, ceramic plants are located close to the mining operation whenever possible. In some cases, the raw material is taken directly from the mine to the plant with little or no treatment at the mine. In other cases, it is processed and refined (beneficiated) near the mine and then shipped to distant ceramic markets by railroad or truck. Feldspars, kaolins, and kyanite are examples of materials that are beneficiated near the mine before shipment.

Crushing and grinding

Ceramic plants that use large quantities of bulk raw materials frequently have a sizeable investment in machinery designed to crush or pulverize raw materials and classify (separate) them according to particle size.

Jaw crushers and gyratory crushers are frequently used as primary crushers to reduce the size of lumps of raw material from as large as a yard in diameter down to a few inches in diameter. The jaw crusher uses a horizontal squeezing or "chewing" motion to fracture the large lumps of raw material between stationary and reciprocating vertical plates of hardened steel (Fig. 3.2). The gyratory crusher uses eccentric rotating motion to accomplish a similar crushing action between concentric steel cones (Fig. 3.3).

Secondary crushing refers to reduction in size from lumps a few

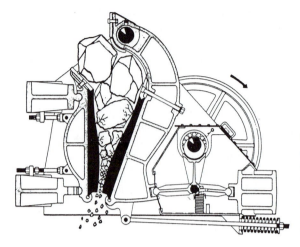

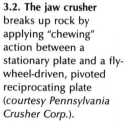

3.2. The jaw crusher breaks up rock by applying "chewing" action between a stationary plate and a fly-wheel-driven, pivoted reciprocating plate (*courtesy Pennsylvania Crusher Corp.*).

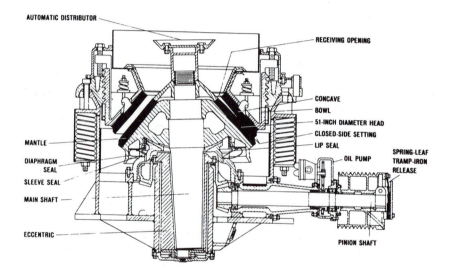

3.3. The gyratory crusher is similar in principle to a jaw crusher, but a stationary inverted bowl and eccentrically gyrating, inner cone are used to apply the "chewing" pressure to the rock (*courtesy Pennsylvania Crusher Corp.*).

inches in diameter down to particles 0.1 inch or less in diameter. Clay materials frequently undergo secondary size reduction by crushing under heavy steel-tired wheels (mullers) in a shallow rotating pan. Plows force the clay under the heavy mullers, and perforations in the bottom of the pan allow material of sufficiently small size to drop through. This crushing device is called a dry pan. After dry panning, many clay materials need no further crushing (Fig. 3.4).

3.4. A dry-pan crusher showing the heavy mullers and plows (*courtesy J.C. Steele & Sons, Inc.*).

Hammer or impact mills may be used for the secondary crushing of hard, brittle materials. These mills break down material inside the chamber by impact with steel hammers mounted on a rapidly rotating shaft. The hammer mill uses a bottom screen to control the maximum size of particles leaving the mill chamber; the impact mill operates like a hammer mill but has no screen and produces a less uniform product. Cone crushers, which operate on the same principle as gyratory crushers, may also be used for secondary crushing of hard materials. Other secondary crushers in common use are smooth-roll crushers, which crush the material by pinching between two narrowly spaced, oppositely rotating rollers, and toothed-roll crushers, which crush material between a toothed roller and a curved steel plate (Fig. 3.5).

When a very fine particle size product is required, as for example in the cement industry or in preparing ceramic casting slips, a tertiary size

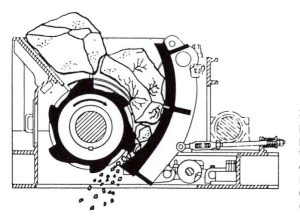

3.5. A toothed-roll crusher pinches rock between a plate and a rotating toothed roll to give a combination of shear, impact, and compression (*courtesy Pennsylvania Crusher Corp.*).

reduction step is necessary. This operation is termed fine grinding and is often accomplished in some kind of a tumbling mill. These mills consist of large, horizontal, rotating cylinders partially filled with heavy steel or dense ceramic grinding media of spherical shape (ball mill, Fig. 3.6), or in the form of heavy rods of the same length as the cylinder (rod mill). Short ceramic cylinders are now used as grinding media in many ball milling operations. Occasionally flint pebbles will be used as the grinding media (pebble mill). Raw material smaller than an inch in diameter is fed to these mills, and the resulting product can be 50 μm or less in diameter, depending on grinding time. During rotation of the mill, this tremendous size reduction takes place by impact of the material with the tumbling grinding media and by abrasion between the media and the mill wall. If no iron contamination of the product can be tolerated, all-ceramic-lined mills and ceramic grinding media must be used. Milling can be done dry, but it is much more efficient when the raw material is in slurry form.

In dry grinding, an air separator (whizzer) can be used to carry away the light, fine product while the heavier, coarse material remains behind for further grinding. Dry grinding is sometimes performed with continuous charging and removal through opposite ends. Wet grinding is usually done as a batch operation with filling and discharge through an opening in the side of the cylinder.

Considerable amounts of tramp iron may be present in raw materials after mining, transportation, and crushing, and magnetic separators can be used to remove most of this iron. The type of separator used depends on whether the material is being processed dry or as a slurry. Permanent bar magnet grids or magnetized conveyor terminal pulleys

3.6. Several large ball mills used in the preparation of materials for porcelain spark-plug insulators (*courtesy Abbe Engineering Co.*).

are useful for dry processing, while electromagnetic mesh "filters" may be used for slurries. Superconductor magnetic separators are needed in critical situations where conventional magnets lack the field intensities needed to remove weakly magnetic materials such as rutile (TiO_2).

Screening

Screens are used by the ceramic industries to separate out particles in a specific size range from a mixture of particle sizes. Screens may be used either to eliminate coarse material (the oversize) or to eliminate material that is too small (the fines or undersize). For example, a single screen operating on the output of a crusher can separate insufficiently crushed material from properly crushed material. When the oversize is returned to the crusher, the operation is called closed circuit crushing. Screens can also be placed one above the other (nested) to produce several narrow-size fractions (cuts) that can be stored separately and later reblended in the desired proportions to insure a uniform product. Small, specially constructed screens called testing sieves are used to measure the particle size distribution of ceramic powders (see Chapter 11).

Screens are normally operated in a sloped position to insure a "pouring" action of material over the screen deck. The screen is usually vibrated mechanically or electromechanically to aid in material flow and improve separation efficiency. Figure 3.7 illustrates vibrating screens.

3.7. Vibrating screens:

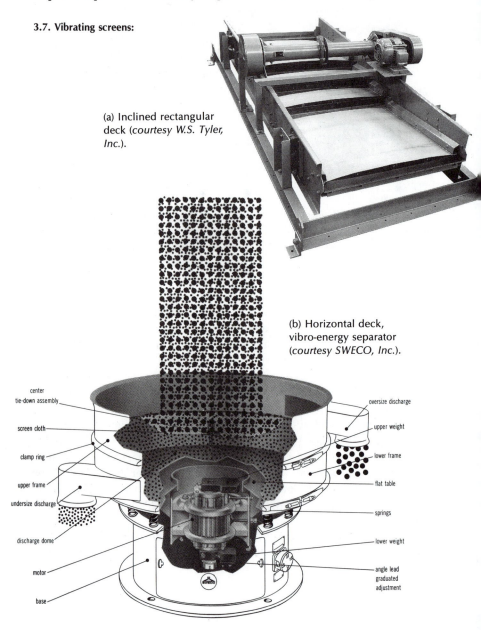

(a) Inclined rectangular deck (*courtesy W.S. Tyler, Inc.*).

(b) Horizontal deck, vibro-energy separator (*courtesy SWECO, Inc.*).

center tie-down assembly

screen cloth

clamp ring

upper frame

undersize discharge

discharge dome

motor

base

oversize discharge

upper weight

lower frame

flat table

springs

lower weight

angle lead graduated adjustment

The size of screened material is normally specified by mesh numbers. The mesh number of a screen is the number of openings per lineal inch of screen surface. The higher the mesh number, the smaller the opening size. The size openings of various mesh screens are given in Table 11.1 in Chapter 11. If material has been passed through an 8-mesh screen, it may be referred to as "minus 8-mesh," "through 8-mesh," or simply "8-mesh." If the material has passed through an 8-mesh screen, but did not pass through a 10-mesh screen, it is referred to as "$-8 + 10$ mesh," "through 8-on-10-mesh" or simply as "8/10" material.

The efficiency of a screen is a measure of its ability to perform the separation or "cut" between oversize and undersize material. An inefficiently operating screen will usually produce an oversize stream that actually contains an appreciable amount of fines that should have fallen through the screen. This will occur whenever a sizeable fraction of the screen openings are covered over or plugged by material. If the screen is fed too rapidly, much of the moving material will never contact the screen surface and thus can never be sized. Plugging of the screen openings (blinding) can be caused by particles only slightly larger than the opening (so-called near-mesh particles) or by sticky materials such as damp clays. The latter problem can be reduced by electrically heating the wire screen to dry out damp material. Wear of the screen by abrasive materials can enlarge the openings so that coarse material enters the undersize stream. This will also constitute a loss in efficiency.

Materials handling

Many ceramic manufacturers purchase large quantities of their raw materials already crushed, screened, and purified. These materials are received either as bulk shipments in railroad hopper cars or in bags transported by railroad car or truck. Bagged materials may be stacked under cover in an ordinary warehouse facility, but bulk materials must be stored in bins or silos, which are discharged through the bottom. The lack of uniform flow of powdered materials in a bin, and especially through the discharge opening, is a cause of many production problems, especially when nonfree-flowing materials like clays are involved. The removal of material from the bottom of a bin is often hampered by formation of impacted material above the opening (bridging or arching) due to the weight of material above. Careful design of bins, including consideration of bin wall slope angles and interior surface finish, can sometimes alleviate flow problems. The most common solution to flow problems in bins, however, is to incorporate some kind of bin vibrator or internal stirring device to improve flow.

Bagged materials are generally transported around a plant by means of forklift trucks, but bulk materials are usually transported by means of conveyors (Fig. 3.8).

3.8. Bulk material conveyors:

(a) Screw or auger conveyor (*courtesy Screw Conveyor Corp.*).

(b) Belt conveyors (*courtesy Stephens-Adamson Division of Borg-Warner Corp.*).

(c) Flight conveyor (*courtesy Jeffrey Manufacturing Co.*).

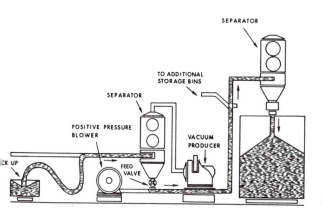

(d) Diagram of a pneumatic conveyor system with vacuum pickup and pressurized distribution to bins (*courtesy Spenser Turbine Co.*).

SEPARATOR

TO ADDITIONAL STORAGE BINS

SEPARATOR

POSITIVE PRESSURE BLOWER

VACUUM PRODUCER

FEED VALVE

CK UP

1

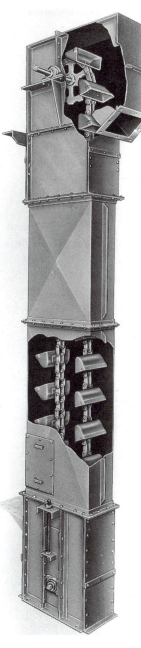

(e) Bucket elevator (*courtesy Stephens-Adamson Divison of Borg-Warner Corp.*).

A screw conveyor has an auger turning inside a trough and is useful for moving nonabrasive, nonfriable materials of fine particle size over short distances up to perhaps 100 feet. Sloping vibratory conveyors are useful for very short distances and are particularly important for "dribbling" a small amount of additive into a blending operation. For medium- or long-distance conveying, endless belt conveyors are widely used either for horizontal or inclined movement of large tonnages of lumpy materials. Flight conveyors consisting of a series of vanes (flights) on an endless chain are sometimes used for medium-distance conveying of materials, especially if they are hot. Pneumatic conveyors that push or pull fine powders through a pipeline by means of differences in air pressure are replacing many other types in the ceramic industries. Many plants today receive dry bulk materials in specially constructed railroad tank cars or tank trucks, which can be unloaded pneumatically in a few minutes. Pneumatic conveyors can move fine material vertically; many other types cannot do so. Bucket elevators are used for moving lumpy material vertically.

A popular concept for kaolin and ball clay users is the purchase of material in the form of a water slurry delivered by railroad tank car or tank truck. Other raw materials can be added directly to this slurry to form a slip for casting purposes (see Chapter 4).

Proportioning and mixing

The success of the modern ceramic industries is based on an ability to correctly proportion various raw materials of the proper purity and particle size distribution, mix them together, form them into useful shapes, and finally fire them into dense strong forms at high temperatures. In the structural clay products industries, the proportioning consists merely of feeding a single pulverized shale into a pug mill at a controlled rate along with a controlled amount of water. In most of the ceramic industries, however, several different raw materials need to be carefully weighed out and mixed together.

If the proportioning is to be done on a batch basis (i.e., if a definite amount of the mixture is to be prepared), the ingredients are weighed one by one into a weigh hopper. The weight of each ingredient added to the hopper is usually automatically recorded. When batching is complete, the weigh hopper is discharged into the mixing machinery. The weigh hopper may be stationary with a conveyor used to transfer the batch to the mixer, or the weigh hopper may be portable so it can be moved under each raw material bin to receive the proper amount of material and then be moved to the mixer.

When proportioning is to be done continuously rather than in discrete batches, variable-speed belt or screw conveyors are used to carry each material from its storage bin into a continuous mixer. The speed of each raw material conveyor is fixed by the relative proportions of the raw material in the recipe and the demands of the process.

Bagged materials can be loaded directly into a batch mixer. In the manufacture of technical ceramics, bags are usually weighed before and after emptying so that the actual weight of material added can be determined. This practice is necessary because small variations in composition of certain technical ceramics can have profound effects on the final properties. Very small additions are carefully weighed on accurate table balances before being added to the batch.

Dry mixing is usually accomplished in either a shell mixer, a ribbon mixer, or an intensive mixer. The twin-shell or "V" mixer shown in Figure 3.9(a) consists of two hollow cylinders joined together into a V shell. As the shell rotates around the horizontal shaft, the tumbling material is divided and poured together again and again to accomplish mixing. Another shell mixer configuration is the double cone shown in Figure 3.9(b).

3.9. Shell mixers:

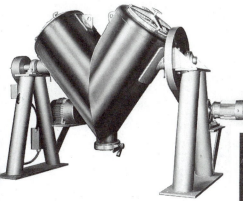

(a) Twin-shell or V mixer. Mixer is charged through the large ends and discharged through the small opening at the bottom of the V (*courtesy Patterson-Kelley Co., Inc.*).

(b) Double-cone mixer (*courtesy Patterson-Ludlow Division of Banner Industries, Inc.*).

The ribbon mixer (Fig. 3.10) uses helical vanes (ribbons) rotating inside a closed chamber to accomplish mixing. Figure 3.11 shows a design called a dry agitator, which utilizes a segmented ribbon.

The intensive (turbine) mixer uses rapidly revolving plows to give coarse material a tumbling and folding action.

3.10. Conventional ribbon mixer (*courtesy J.H. Day Co.*).

3.11. Dry color agitator with a segmented ribbon (*courtesy J.H. Day Co.*).

Tempering

Tempering is the addition of water or other liquids to a dry ceramic powder to produce a mixture suitable for forming. The tempering operation usually combines mixing of dry materials with the incorporation of water by repeatedly cutting and kneading the mass. The result can be material with a few percent moisture content useful for dry press forming or can be a paste suitable for plastic forming (see Chapter 4 for discussion of forming procedures).

Shales and fire clays that are to be formed by extrusion are usually tempered in a pug mill (Fig. 3.12). The machine uses angled knives on a rotating shaft to cut through and fold together material as it slowly moves along the shallow chamber. Pulverized clay and water are usually fed more or less continuously to one end of the pug mill, and the tempered mass exits at the other end, usually into an extrusion press.

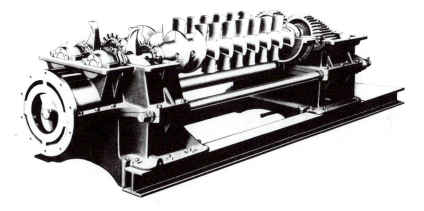

3.12. Pug mill with exterior shell removed to reveal pugging assembly (*courtesy Fate-Root-Heath Co.*). See also, Figure 4.1(a).

Sigma blade mixers (Fig. 3.13) can give effective mixing of several dry materials in addition to water incorporation. Two oppositely rotating S blades force the powder and water together or against the steel body of the mixer in a kneading action.

3.13. Sigma-blade mixer (*courtesy J.H. Day Co.*):

(a) Mixing position.

(b) Tilted in unloading position showing the mixer blades.

The mix-muller (Fig. 3.14) consists of revolving rolls or mullers on an axle extending from a rotating pivot. The rolls ride in a stationary, shallow circular pan where plows guide material under them for tempering. Discharge of the batch of tempered material is through a gated opening in the side of the pan. Such a mixer is sometimes called a wet pan (note similarity to the dry pan crusher).

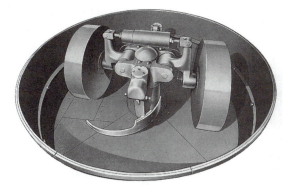

3.14. Mix-muller (*courtesy National Engineering Co.*):

(a) Details of construction showing mullers and plow.

(b) Mixing a tempered material showing the smearing action of the mullers.

Both intensive mixers and shell blenders can also be used to add very small amounts of moisture to a ceramic mixture. With a shell blender, water can be sprayed into the mix through the perforated hollow shaft on which the mixer turns.

Slurry processing

The manufacturers of whitewares and technical ceramics often find it necessary to add enough water to their raw material mixtures to form a slurry or *slip*. One reason for adding so much water is to aid in the mixing of a multiple-ingredient batch. A second reason for preparing such a slurry is to permit use of the slip casting method for forming ware (see Chapter 4).

Frequently, slips are prepared by combining the mixing, fine grinding, and water addition steps all in one operation. The ball mill is a useful device for accomplishing these objectives. A vibrating mill, shown in Figure 3.15, is also capable of combining slurry mixing and fine grinding. This mill relies on the rubbing action of a bed of continuously vibrating short ceramic cylinders to accomplish the grinding and mixing. Fast grinding with little contamination is a major advantage of the virbrating mill. The attrition mill (Fig. 3.16) utilizes relatively fine grind-

3.15. Vibro-energy mill
(*courtesy SWECO, Inc.*):

(a) Wet grinding mill.

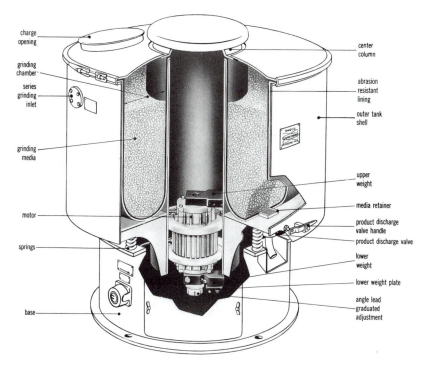

charge opening
grinding chamber
series grinding inlet
grinding media
motor
springs
base

center column
abrasion resistant lining
outer tank shell
upper weight
media retainer
product discharge valve handle
product discharge valve
lower weight
lower weight plate
angle lead graduated adjustment

(b) Schematic.

ing media inside a small sealed chamber to increase grinding efficiency further with tighter particle size distributions. Rotating agitators move rapidly through the mixture of media and slurry, giving tne approximate equivalent action of a ball mill.

Dry materials of sufficiently fine grain size can be made directly into slips by mixing with water in *blungers,* which are large wooden or metal tanks with paddles or turbines used for agitation. The tanks are usually baffled to eliminate vortexing of the mixing fluid. After blunging, slips are usually stored in slowly agitated holding tanks to prevent settling.

Because of the abrasive nature of ceramic slurries, they can be pumped only with centrifugal or diaphragm pumps that do not have close fitting parts. Transportation from holding tank to utilization point in the plant may be carried out using a pipeline or using small portable tanks equipped with a pump for emptying.

If excess water has been added only to aid grinding or mixing, it is usually necessary to remove part or all of the water before forming. Partial dewatering can be accomplished by forcing the slip through a

3.16. Attrition mills
(*courtesy Netzsch, Inc.*):

(a) Horizontal attrition
mill. Slurry is milled in
the sealed chamber at
right.

(b) Vertical attrition mill.
Slurry is milled in a
chamber wheeled into
position.

filter press such as that shown in Figure 3.17. These filters remove some
of the water and leave behind a plastic mass (filter cake) that can be
extruded for plastic forming or can be shredded for further drying.

3.17. Filter press (*courtesy Gebrueder Netzsch*).

A particularly useful device for reducing a slurry to a granulated product suitable for dry forming is the spray drier (Fig. 3.18[a]). In the configuration illustrated schematically in Figure 3.18(b), the slip is pumped to an atomizer consisting of either a rapidly rotating disk or a nozzle located inside the top of the spray drier. Droplets of slip are thrown out toward the edge of the chamber as they fall through a rising stream of hot air, drying completely before they hit the walls or bottom of the drier chamber. The resulting product consists of small, free-flowing granules.

3.18. Spray drier for dewatering slurries (*courtesy Bowen Engineering, Inc.*):

(a) Drier chamber with air heater and cyclone separator.

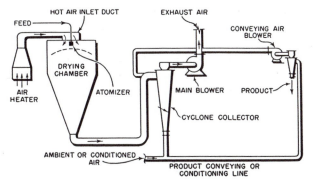

(b) Schematic of a typical spray drier.

The shape and size distribution of spray-dried granules depends on the particular spray drier, the type of atomization, the properties of the slip, the type and amount of organic binder added to the slip, the temperature of the drier, and the nature of the raw material. The dried product is usually removed at the bottom of the drier and may be placed in portable hoppers or in large standardizing tanks which use forced air to mix together several batches from the spray drier to insure uniformity. The use of spray-dried material in press-forming operations is discussed in the next chapter.

Synthesis of materials for advanced ceramics

Ceramics used in advanced applications must be especially carefully tailored to the final property requirements of the particular application. This process begins with the synthesis of materials of the correct composition, mineralogical structure, particle size distribution, particle shape, and especially purity. Changes that take place during subsequent processing of these materials will not correct for deficiencies occurring during the initial synthesis.

Chemical synthesis

Since World War II there has been great interest in making ceramic powders of very high purity from pure chemical *precursors* — starting substances from which other substances are formed. For example, aluminum oxide can be easily made in the laboratory by dissolving high-purity aluminum metal in hydrochloric acid, precipitating the aluminum as aluminum oxalate by adding ammonium oxalate, filtering and drying the aluminum oxalate, and calcination to the pure oxide at a controlled temperature for a specified time. The final product purity, reactivity, and particle size distribution will vary with the processing parameters and procedures.

Fairly high purity aluminum oxide has long been manufactured in extremely large quantities for use in ceramics by chemical processing. The process (called the *Bayer process*) begins with mined bauxite, which is digested in a hot pressurized solution of sodium hydroxide, followed by crystallization of aluminum from solution as aluminum hydroxide, which is subsequently calcined to oxide. During calcination, addition of a small amount of boron allows sodium to be removed as volatile sodium tetraborate. This type of process does not yield the very high purity material sometimes required in advanced applications, because the starting material (impure bauxite) usually results in some retained impurities. However, Bayer process alumina is satisfactory for spark plug insulators, which must face high mechanical, thermal, and corrosive environments; for refractories used in steel making where the temperatures are near 1600°C (2912°F) and corrosive molten slags and metals are present; and for missile nose cones (radomes), which must maintain critical electronic and mechanical parameters under adverse missile flight conditions.

Coprecipitation is a chemical process in which solutions of pure soluble salts of two or more metallic ions are blended allowing very intimate mixing. While still in solution, chemical processing techniques may be used to further purify the components before the precipitation process begins. On precipitation, usually by adding a desired anion, an intimately mixed precursor material results. This material may be calcined to a multicomponent oxide under controlled conditions, resulting in a high-purity ceramic-grade powder. A high degree of process control is required with due regard for chemical solubility products. Important parameters are solution pH, concentration, temperature, and order of mixing. Steps must be taken to avoid contamination of the refined powder during calcination. Sometimes refractory containers of the same composition as the desired powder can be used to preserve purity during calcination.

An example of this chemical route is the formation of very pure spinel, $MgO \cdot Al_2O_3$. Magnesium and aluminum chloride hydrates are dissolved in pure water. When the solution is made basic to a pH of 9.5, say, by adding NH_4OH, then $Mg(OH)_2$ and $Al(OH)_3$ are precipitated simultaneously. Subsequent calcination and firing will result in a final product that is more uniform than similar products prepared by mixing and firing of MgO and and Al_2O_3 powders. Coprecipitation can produce spinel materials with purity as high as 99.995%. Pure and doped barium titanates can also be produced with this method. These materials are precipitated using the oxalate anion. The oxalate is easily decomposed to the oxide by heating.

Sol-gel is also a precipitation process. Typically metal-alkoxides dissolved in organic liquids are used as starting materials. The chemical form for these materials is $M(OR)_x$ where M is a metal such as silicon, aluminum, zirconium, titanium, or boron, and R is the organic radical. When water is added to the organic solution, the alkoxide groups are removed by hydrolysis and replaced with the hydroxyl group. This reaction is usually catalyzed using acids or bases. The reaction ends with a gelatinelike mass (i.e., a *gel*) made up of a polymerlike precipitate network that is formed by metal-oxygen-metal (M-O-M) bonds. If the gel is carefully dried, it will retain its shape allowing formation of films, fibers, or even monolithic parts. More conventionally, the gel is dried and ground into a fine powder for conventional forming.

Hydrothermal synthesis allows the precipitation of homogeneous oxide powders from pressurized hot water solutions. The high-pressure hot water dissolves the starting material, which would not be soluble in cold water, and new, more desirable crystalline forms can then be precipitated from the solution. Many natural minerals are formed this way underground. The process can be performed in continuous or batch modes. Precursors can consist of oxides, salts, hydroxides, organics, gels, acids, or bases. Lower-cost starting materials having lower chemical purity than required for other chemical processes can be used. Temperature and pressure are controlled in the reaction vessel to drive the desired reactions at rates suitable for producing a material with the desired powder morphology and chemical composition. Water actually acts as a catalyst to the process. The water temperature is usually in the range of 100°C (212°F) to 374°C (705°F) and the pressure is typically 15 MPa or less. After pressure release, the resulting crystallized powder is recovered by filtration and drying. The major benefit of hydrothermal processing is that a high-purity, highly reactive powder is produced without calcining or milling. The process has been demonstrated at production rates up to 100 tons per year.

Plasma synthesis

A commercial technology for synthesis of ceramic coatings and powders, especially for carbides, oxides, and nitrides, is based on reaction within a hot plasma. A plasma is produced when a gas is excited to a state where the molecules separate into individual atoms and then is further excited until a significant fraction of the atoms become ionized. Heat, radiation, or electric discharge may be used to sustain the plasma. To obtain a reasonable degree of electrical conductivity, a plasma must reach a temperature of at least 4000°C (7232°F) and may be much hotter. Ceramic powders can be produced by passing either a solution or a powder-bearing gas stream through the plasma. An example is the refining of zircon ($ZrSiO_4$) powder resulting in zirconium oxide (ZrO_2) with the silica being removed as vapor. There is virtually no contamination introduced by a plasma to the product as there is with more traditional methods where materials are contained by refractory vessels during heat processing.

Plasma processing of materials can be expected to increase in the near future. Research in gas phase reactions to form powders of extremely high purity and fine particle size using high-powered laser systems looks quite promising. Currently special grades of titanium aluminide and titanium diboride are produced by plasma processing of gases. In the production of titanium diboride, titanium tetrachloride and boron tetrachloride are mixed in a high-voltage arc-stimulated hydrogen plasma, producing titanium diboride powder and hydrogen chloride gas.

Carbon thermal reduction

Carbides can be synthesized by heating a metal oxide–carbon mixture, usually by passing a strong electric current through the mass. Part of the carbon removes the oxygen from the oxide as carbon monoxide gas, and additional carbon reacts with the resulting metal to form the carbide. This process, mentioned in Chapter 2 as the traditional method of manufacturing SiC, is becoming more important for advanced materials. An interesting variation of this process is the production of long SiC fibers from precursor fibers formed from a resin copolymer consisting of a silicone and an organic polymer. Upon heating, the carbon from the organic polymer reacts with the silicon in the silicone polymer to produce SiC and gaseous products.

4 Forming

THE ULTIMATE GOAL of all the initial processing steps discussed in the previous chapter is the eventual production of ceramic articles, whether traditional or advanced, that will have properties tailored to a specific end use. To accomplish this goal, the ceramic manufacturer is faced with the task of transforming loose ceramic powders into solid, useful shapes. In some cases such as hot pressing, which is accomplished at high temperatures, the product is completely densified during the forming process, but usually the formed product is later densified by a firing operation. After a ceramic product has been formed, it is often referred to as *ware.*

Ceramics for advanced applications require special care during the forming procedures to assure that a minimum of defects (especially microcracks, voids, and surface pits) are created during forming. Traditional forming methods may be used, but they must be tailored to the materials being processed.

Ceramic parts are more difficult to form than metal parts because of the inherent brittleness, hardness, and high melting temperatures of ceramics in general. Over the years a great variety of forming techniques have been developed for ceramics, and the more common of these will be discussed in this chapter. These techniques are grouped into the categories of plastic forming (sometimes called shear form-

ceramic ware

ing) processes, slip casting, dry forming processes, and high-temperature forming processes. The forming of glass products is discussed in a separate section.

Plastic forming

The ease with which metals can be rolled, drawn, forged, and bent into intricate shapes results from the inherent ductility of the individual crystals making up the metal mass. Each individual crystal can deform under applied pressure and retain its new shape when the pressure is removed. Ceramic crystals are brittle and will usually fracture rather than flow under applied pressures. For this reason a dense mass of ceramic crystals cannot be formed by the same methods used for forming metals.

Prehistoric peoples discovered that clay-containing mineral mixtures tempered with water were plastic enough to be formed by hand into useful shapes. The forming of such plastic mixtures can be accomplished by methods similar to those used for very-low-yield-point ductile materials. The ability to be formed when in the plastic condition and then to be made hard and durable by subsequent drying and heat treatment is responsible for the great utility of clay-based ceramics in our everyday lives.

Today it is known that powdered nonclay ceramic materials, although not rendered plastic by the simple addition of water, can often be brought into a plastic condition by the addition of certain water-soluble organic materials. These additives may be classified as binders or lubricants according to the functions they serve in the plasticizing process. Binders serve to give wet strength during forming and dry strength after drying. Lubricants lower the frictional forces present during forming, permitting forming at lower pressures. Water promotes the plasticizing abilities of these organic materials. Methylcellulose, polyvinyl alcohols, water-soluble gums, starches, proteins, and other organic materials are binders that may be used in the presence of water. Waxes, stearates, and water-soluble oils are examples of lubricants that may be used in the presence of water.

In forming operations carried out above the boiling point of water, the plasticizing abilities of natural or synthetic thermoplastic resins may be promoted with organic solvents called plasticizers. (See section on injection molding.) The resin may act as a lubricant as well as a binder. The organic solvents may also act as lubricants.

Proper selection of plasticizing materials is based on the compatibility of the various organic materials with each other and with the particular ceramic or ceramic-water system. The ability of individual organic materials to satisfy the various plasticizing functions and the ease of removing these plasticizing additives after the forming operation are also key factors. The amount and type of inorganic impurities present in organic additives is an important consideration, because permanent impurities introduced from organic binders can compromise the properties of the final product and can completely nullify the benefit of high-purity ceramic starting materials. The cost and possible toxicity of plasticizing materials are other important considerations.

The plasticizing step—tempering of the ceramic powder with water and (or) mixing with organic additives—is critical to the success of any plastic forming operation. Since the water and other plasticizing additives must be removed prior to firing the ware, only enough are used to develop the minimum plasticity required for the particular forming process. Too much water or organic material may result in warping and excessive shrinkage when they are removed during drying; too little will result in a tendency of parts to crumble and the need for excessively high forming pressures.

Organic plasticizing materials become a critical consideration in technical and advanced ceramics because, on burnout, they can leave behind defects such as pores that may not be removed during firing. These defects can reduce strength, change electrical properties, and in general make the final product useless for critical applications.

The terms *stiff mud* and *soft mud* are applied to plasticized ceramic mixes depending on whether they require a fairly high pressure to cause them to flow. Technically speaking, the terms *high yield point* and *low yield point* can also be used to describe stiff and soft mud mixes respectively.

Extrusion

Extrusion forming of ceramics is accomplished by compacting a plastic mass in a high-pressure chamber (cylinder) and forcing the mass to flow out of the chamber through a specially shaped die orifice. The cross-section of the extruded column (often called the *pug*) takes the shape of the die. Parts of desired length are formed by cutting the extruded column with rotating knives or stiff wires. Obviously, only shapes having a constant cross-section can be formed by extrusion.

The stiff mud forming of most building brick and much refractory brick is accomplished by extrusion, using a clay-water mass of fairly high yield point. Clay and tempering water are continuously fed into a pug mill attached to the extruder (Fig. 4.1[a]). The pug mill cuts and kneads the material to temper it and moves it forward through a shredder into the extruder proper (sometimes called the press). The extruder consists of an auger that moves the tempered mixture consecutively through two sealed chambers. The first or de-airing chamber uses a vacuum to remove air from the shredded clay. The de-airing step is a critical one, since the presence of air pockets in the extruded column would lead to delamination and cracking of the ware. The de-aired clay is then passed by the auger into the second or compacting chamber where the shreds are reconsolidated into a dense plastic mass. The forming die is at the end of the compacting chamber and the auger pushes the plastic mass out through the die where it can be cut to length by a wire cutter shown in Figure 4.1(b). Hollow shapes such as structural tile, flue tile, drain tile, and sewer pipe can be formed by extruding stiff mud through a special die (Fig. 4.2) that contains a central structure made up of a solid core suspended by thin radial supports (the *spider*) within the entrance from the compacting chamber. Sewer pipe production by extrusion is shown in Figure 4.3. (Note that the holes in common building brick are formed by including cores within the rectangular cross-section extrusion die.)

Long, single-piece, high-voltage electrical insulators are often formed by lathe turning of solid porcelain cylinders that have been extruded and then dried.

Extrusion is often used as a preliminary step to other plastic forming procedures. This is especially true of porcelain mineral mixtures that are generally blunged with excess water, filter-pressed to the proper moisture content for plastic forming, and then extruded and wire cut

4.1. Low-pressure extrusion equipment (*courtesy J.C. Steele & Sons, Inc.*):

(a) Auger extruder.

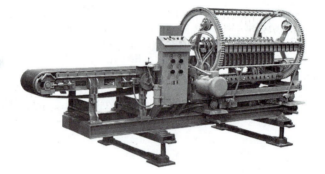

(b) Wire cutter. Wires are stretched across a rotating frame that moves forward in synchronization with the moving extruded column to give a square cut.

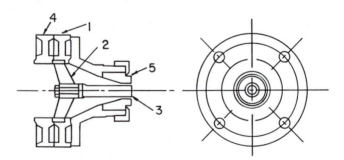

4.2. Schematic of an extrusion-forming die (*courtesy Fate-Root-Heath Co.*). (1) bridge ring; (2) core bridge (spider); (3) core (spindle); (4) extending ring; (5) die.

4.3. Sewer pipe production. The bell-end-forming die is still in
the last extruded piece (*courtesy INTERPACE Corp.*).

into short slugs (sometimes called blanks) that can be fed to other forming machines.

Nonplastic high-yield-point ceramic mixes are usually extruded
from a vertical or horizontal piston press rather than with an auger.
These materials can sometimes be extruded at low pressures, but high-
pressure extrusion reduces the amounts of binder and moisture required
during extrusion and yields a denser and more uniform final product. A
common practice in plasticizing nonclay ceramic mixes is to first blend
the various dry ceramic raw materials together in a sigma-blade or mul-
ler mixer. Dry plasticizing additives may be added before water additions
or they may be added in the form of water suspensions. For high-pressure extrusions, the moisture content is usually reduced to the point
where the material tends to form granules about 1/4 inch in diameter in
the mixer. These granules are easy to pour into the extrusion press.
Extrusion presses capable of exerting more than 100 tons are commer-

cially available, but presses of 30 tons capacity are more common. Figure 4.4 shows a typical piston extrusion press. The extrusion cylinder is usually made of tough, wear- and corrosion-resistant alloy steel. Dies may be fabricated to extrude shapes with a variety of cross-sections. Oxide thermocouple tubing, insulators, rods, and protection tubes are typical of nonplastic ceramic shapes produced by extrusion. Shapes with critical tolerances can be partially microwave-dried as they leave the extrusion head by passing them through a coil radiating at microwave frequencies.

4.4. High-pressure extrusion press with adjustable tilting extrusion cylinder (*courtesy Midvale-Heppenstall Co., Pressure Equipment Division*).

Plastic re-pressing

In the refractories industry, stiff extruded clay is cut into bricks or special shapes, which are then re-pressed to final shape in steel dies before drying. An improved part results from the re-pressing due to higher density and better dimensional uniformity and edge integrity.

Wet pressing

In wet pressing, a soft mud plastic mass containing clay as a major constituent is formed by pressing a de-aired and extruded blank or shredded granules into a steel or plaster die. As pressure is applied, the plastic mass flows into the various contours of the die. Typical products formed in this way are electrical insulators and special refractory shapes. Figure 4.5 shows a typical press and die assembly which utilizes plastic granules.

4.5. Wet press forming perforated discs. Granules flow into the die (center) through the "shoe" (left) as it slides over the die opening and back. The bottom punch ejects the formed part, which is moved by the sliding "shoe" onto the conveyor belt (right) as a new charge of granules falls into the die cavity (*courtesy AC Compacting Presses, Inc. and Dorst American, Inc.*).

In one version of the process, the steel die is heated. (This process is sometimes called "hot pressing," but confusion between this process and conventional hot pressing is best avoided by considering it to be simply a modification of normal wet pressing.) Heating causes a steam cushion to develop between the wet plastic mass and the die surface, acting as a lubricant to promote easy removal of the formed part. If the die is not heated, it must be lubricated. In some operations the die is rotated. Presses are now available that allow the automatic production of small parts. As in all plastic processing, the moisture content of the body must be carefully controlled. Proper de-airing of the starting material is also an essential step.

Jiggering

The *jiggering* process forms a de-aired and extruded soft mud blank to the contours of a rotating plaster mold with the aid of a hardened molding template, which forms the contour of the side opposite to the plaster mold. The process is limited to round (sometimes oval) shapes made from ceramic mixes containing significant amounts of clay. This process is especially suited to the manufacture of whitewares. Chemical porcelain and china dinnerware are typical products. In a manual jiggering operation, a skilled operator wets the plaster mold, places the extruded slug onto the mold, and uses the template to press the plastic body carefully out to the contour of the rotating plaster mold. The template is forced down until the final inner contour is formed. The operator removes the excess clay as it is sheared away by the template, raises the template, and finishes (*fettles*) the piece with a sponge.

The jiggering process is almost always automated and the rigid forming template is often replaced with a shaped roller (roller forming). Figure 4.6 shows an automatic jiggering machine.

High losses, usually during drying, can occur if the blend of materials for plastic jiggering is not properly selected and prepared. Improper control during blunging and filter pressing will lead to serious problems. The very fine particles in the blunge slip must be chemically aggregated

4.6. Automatic jigger forming several different shapes at once (*courtesy Syracuse China Corp.*).

from their dispersed conditions into what is called the *flocculated* state before filter pressing and extrusion prior to final forming.

In the forming of some special fine china items and cups, the *cast jiggering* process is often used. Rather than starting with an extruded clay blank, slip is poured into the plaster jiggering mold, allowed to set to form the handle and a thin deposit on the mold surface, then the inside surface is jiggered. In this case, the clay slip is dispersed rather than flocculated. Cups are also formed by conventional plastic roller forming, with the handles being separately formed and then stuck to the undried cup proper using slip to act like a glue.

Many manufacturers of dinnerware have converted or are converting from jiggering to isostatic (dry) pressing of flat dinnerware shapes and shallow bowls. This process is described later in this chapter.

Ram pressing

In the *Ram process,* de-aired, extruded soft mud slugs are pressed between two halves of a special hard plaster mold mounted in a hydraulic press. As the mold halves close, the plastic mass fills the mold cavity and the excess extrudes out the sides, forming a flashing. Air is forced through the bottom half of the porous mold by means of copper tubing embedded in the plaster, while vacuum is pulled on the top half of the mold so that, when the top half of the mold is raised, the formed piece separates from the bottom half and is raised with it. Air is then forced through the top half, causing the formed part to separate from the mold. The flashing is removed from the piece, and an additional slight amount of machine or hand finishing may be needed. This process is particularly well adapted to the forming of most shallow shapes. Figure 4.7 shows a Ram pressing operation.

4.7. Ram pressing. At the end of the cycle, air pressure on the top half of the mold releases the part into the operator's hands (*courtesy Haeger Potteries, Inc.*).

Fired porous ceramic molds have replaced plaster molds in many plastic pressing operations. Such molds are expensive but this is justified for large production runs (up to 50,000 pieces) by their much longer mold life. Alumina compositions are usually used.

Plastic pressing of clay slugs in porous plaster or ceramic molds is being replaced by high-pressure slip casting in some sectors of the white-wares industry (see Slip casting).

Injection molding

In the traditional injection molding process, dry nonplastic ceramic powders are first mixed with resin binders and special plasticizers. The mixing is usually done in steam-jacketed mixers at temperatures above the softening point of the resin so that the mix attains a doughlike consistency. Many resins can be obtained in water emulsion form, to be added to ceramic mixes at room temperature, then dried before plasticizing. Special resins are now customized by chemical manufacturers to optimize the process. More recently, water-based gel systems have been developed that yield a more uniform defect-free product as required in advanced ceramic applications.

The heated, plasticized material is formed into small pellets by extrusion through a multiple-orifice die. In the injection molding process (Fig. 4.8), the pelletized mix is loaded into a heated injection molding cylinder, where it softens and a steel piston moved by an air-actuated cylinder rapidly forces the hot plastic mixture into a cooled metal mold. The part sets quickly as the resin binder cools. The mold is opened and the rigid part is removed and broken away from the sprue and any flashing that developed during molding. This forming method is especially suited to the fabrication of small parts with intricate shapes. Injection molding is also used to manufacture special coarse-grained refractory shapes using paraffin binders which bleed out into a supporting bed of alumina sand during the first stages of firing.

Resins and plasticizers must be completely removed from the parts before firing. This is accomplished by a carefully controlled drying cycle, often under reduced pressure or other special conditions. Most synthetic resins and special plasticizers can be obtained in a range of molecular weights, which allows control of the melting range. Resins and special plasticizers of different molecular weights can be blended if they are compatible. On heating, the low molecular weight materials evaporate first, providing a pore network to facilitate the removal of the higher molecular weight materials as the temperature is increased. If at least one resin has a slightly thermosetting character it will help retain the dimensional stability of the part during binder removal. In this age

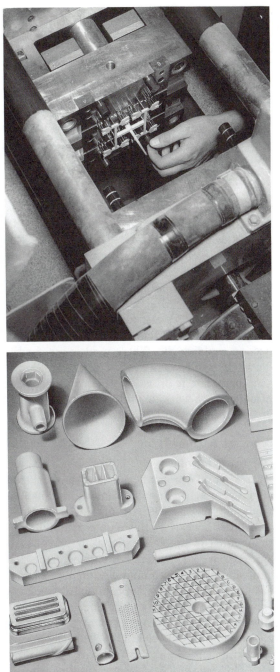

4.8. Injection molding of ceramics (*courtesy American Lava Corp.*):

(a) Removing parts from opened mold. The heated injection cylinder can be seen at the lower left.

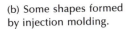

(b) Some shapes formed by injection molding.

of computer chemistry and special molecular synthesis, polymers for use in injection molding of ceramics are now tailored by chemical companies to the particular product being manufactured.

There is major emphasis today placed on water-soluble binder systems. The natural gum agar that has long been used as a stabilizing and gelling agent in foods is of much interest. Water-based systems are preferred when plastic ceramic ingredients like clays are included in the formulation of the ceramic body. Binder removal is less complex and generally less detrimental to the final product if water-based systems are used. The process reduces to simple drying.

Gel casting

Gel binder systems, some of which are related to injection molding formulations, may be cast in almost any kind of mold, porous or nonporous. An activator is used to cause the slurry to set into a gel after pouring. The main ingredients of a gel cast system are the ceramic powder, water (the solvent), cross-linkable organic monomers, dispersant, initiator, and catalyst. As with injection molding, a binder burnout cycle is required prior to firing.

Models for complex shapes to be gel cast can be injection molded with low-melting or water-soluble waxes if the production volume warrants the cost of the injection molding die. These models are used to make working molds. The model can be easily removed from the working mold by melting or by dissolving it in water. After the model has been removed, the cast is made in the working mold. The working mold may have to be destroyed to recover the formed part if the shape is complex.

Tape casting or doctor blade forming

Thin ceramic parts of intricate shape can be machine stamped from predried sheets of plasticized ceramic materials. Forming the material initially into a sheet avoids uneven particle size distributions during forming, gives excellent thickness control of the final parts, and eliminates many problems encountered in the conventional dry pressing of small thin parts. The sheets from which parts are to be stamped are usually first formed from a slip into a continuous flexible tape, and so the process is often called *tape casting*. The appropriately mixed and sized ceramic powder is made into a slip by ball milling in an organic solvent containing a large amount of dissolved organic resins which will later serve as binders and plasticizers. After bubbles have been removed from the slip by vacuum de-airing, the slip is allowed to slowly flow out from a reservoir onto a moving smooth flat surface, passing under a hard rubber "doctor blade" suspended parallel to and just above the moving surface. The doctor blade spreads out the slip onto the moving surface into a thin layer of uniform thickness. The separation distance of the blade from the surface fixes the thickness of the slip deposit.

The surface onto which the deposit is bladed is usually a tightly stretched thin mylar sheet fed from a roller just upstream of the slip reservoir. This mylar sheet serves as a temporary support and backing until the slip deposit is cured. Curing consists of removing the organic solvent, leaving behind the dissolved resins as a binder. The cured material is stripped away from the mylar backing and remains as a thin, pliable tape which can be rolled up and stored or can immediately be cut into sheets from which thin parts can be stamped. Since the tape contains large amounts of organic resins, it is necessary to carefully remove these by an oxidizing burnout prior to firing. The large amount of porosity remaining after resin burnout means that the parts will show a high firing shrinkage as densification takes place.

The tape casting process is especially suitable to forming the extremely thin sheets of insulating ceramics used as substrates for thick film hybrid circuit and chip packages for electronic circuitry. In an interesting variation of the process, patterns of conducting metal electrodes can be screen-printed onto one or both sides of ceramic tapes, and several layers of tape can be stacked on top of one another, laminated together by warm pressing, then fired to form so-called monolithic multilayer ceramic (MLC) structures. This is an especially important technology in making complex ceramic electronic packages and the popular multilayer ceramic chip capacitors.

Slip casting

The dewatering of a slurry by filtration with a gradual buildup of filter cake in contact with the filter surface is the principle underlying *slip casting*. A slip is a slurry of very fine ceramic materials suspended in water (or occasionally another liquid) and generally having the consistency of thick cream. If such a slip is poured into a porous plaster mold, the mold will draw water from the slurry and will build up a deposit of ceramic particles on the mold wall. In this manner a ceramic article (*cast*) can be formed having an outer configuration that reproduces the inner configuration of the plaster mold.

Wall buildup during slip casting continues as long as any slip remains in the mold. However, the rate of buildup drops off with time, because water can be removed from the slip only by passing through the thickening cast wall. If a hollow cast is desired, the excess slip is poured or drained out of the mold after the required wall thickness has been achieved, and for this reason the operation is sometimes called *drain casting*. It is also possible to cast solid objects of small cross-section if a large enough slip reservoir is designed into the mold. Figure 4.9 shows the operations performed to produce hollow or solid casts (see also Fig. 1.3[b]).

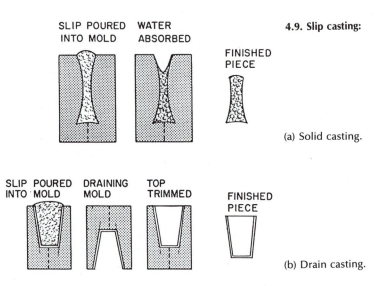

4.9. Slip casting:

(a) Solid casting.

(b) Drain casting.

The preparation and care of plaster molds is critical to the success of a slip casting operation. Suppliers of plaster for the ceramic industry are well informed in this area, and specific information is easily obtained from them. The amount of setting water used in mixing the plaster determines the strength and porosity of the resulting molds. Since the strength of plaster decreases and the porosity increases with an increase in the amount of setting water, the mixing ratio of plaster to water is always a compromise. From 68 to 90 parts by weight of water to 100 parts plaster are used in the ceramic industry. Only potable setting water should be used for plaster molds, and they must be dried carefully before and between uses. Plaster is such a poor conductor of heat that molds may crack if subjected to a sudden temperature change.

Ceramic articles are generally cast in molds made up of several close-fitting parts. The mold is disassembled after casting to remove the cast. The mold parts must be tightly held together during casting, as a significant hydrostatic pressure builds up in the mold due to the weight of the slip. If not held tightly closed, the slip will run out between the joints of the mold. Re-entrant angles (undercuts) cannot be tolerated in single-piece plaster molds since the cast thus formed could not be removed from the mold.

The flow properties and, more importantly, the casting behavior of ceramic slips depend strongly on slip specific gravity and viscosity, and most importantly, how the viscosity changes with time (gelling behavior). The specific gravity is simply a measure of the amount of solids suspended in a given volume of slip. In general, the specific gravity of slip is kept as high as possible, consistent with proper viscosity parameters. The viscosity of the slip is somewhat a function of the mean particle size of the solids present, but is more strongly dependent on the nature of the solids present. The flow properties of casting slips are generally controlled by means of small amounts of dispersion additives, defloculants, so-called because they break up agglomerates called "flocs." Common deflocculants used with slips containing clays are sodium silicate, sodium carbonate, sodium phosphate, and a number of organic substances such as the polyacrylates. For nonclay slips, strong acids or strong bases often give dispersion, but certain organic substances are also suitable for this purpose.

If the particle size distribution of a slip is too coarse, the slip will be difficult to disperse and will settle out rapidly, giving uneven casts. If the size distribution is too fine, the slip will cast too slowly. Aging of freshly prepared clay slips often improves their casting behavior, but the proper aging time, as well as specific gravity, temperature, and viscosity for proper casting must usually be determined by experiment.

Nonclay ceramics are usually more difficult to cast than conventional porcelains. Dense, fused oxides cast more easily than fluffy precipitated oxides. Materials such as MgO and CaO which react with water may be cast from a suspension in absolute alcohol or other inert liquid. A low slip temperature (35°F, 2°C) inhibits hydration in aqueous MgO casting slips.

When a cast is dried, it shrinks free of the mold and can be removed for finishing and firing. Casts containing clays usually have high green (unfired) strength and can be readily handled prior to firing. Casts of many nonclay ceramics are extremely delicate and must be handled with great care before firing.

Uniaxial casting is a process sometimes used to form complex shapes, with an impermeable hollow disposable mold sitting on a flat plaster base. (The models for these molds may be formed by injection molding of water-soluble or low-melting waxes which can be easily removed from the final working mold. The mold is simply heated to allow the melted model to run out, or the model is dissolved out.) In uniaxial casting, conventional casting slip is poured into the impermeable mold and water is drawn out through the plaster base. Casting times are long because the only water removal must occur through the gradually thickening part. Gel casting reduces the casting time and gives a more uniform grain structure.

Pressure casting

Slip casting with pressure applied to the slip in the mold was of limited use until the recent invention of strong nonplaster synthetic polymer molds. This process has particularly improved the productivity of large sanitary ware shapes and whiteware products for the domestic market. Completely automatic production lines are in operation. Dow Chemical Company has shown considerable interest in the pressure casting process and has developed dispersants and casting aids that dramatically increase productivity. A conventional casting slip must be adjusted to make it suitable for pressure casting operations, with particle size distribution and slip specific gravity being key parameters to control. Better processing efficiencies can often be achieved by reformulation of conventional casting slips, replacing some of the plastic (clay) ingredients with nonplastic ingredients.

Dry forming

Ceramic ware can be formed under high pressure from powders with relatively low moisture content or often with no moisture at all. In the latter case, organic lubricants and binders are often utilized. Dry forming of ceramics implies that the material contains 4% or less water. Spray-dried powders are particularly suitable for dry forming processes.

Dry pressing

Dry pressing is accomplished by compacting powders under very high pressures in steel dies and is particularly well suited to the rapid production of a great variety of ceramic parts. Sometimes the forming of mixes containing up to 4% moisture is called "dust-pressing" or "semidry pressing," and only the forming of bone-dry mixes is called "dry pressing," a distinction that is not made in this book. In a typical dry pressing operation, freely flowing ceramic granules containing organic binder or a low percentage of moisture flow from a hopper into the forming die. The granules are typically agglomerates which break down during pressing. The material is compressed in the hardened steel cavity by steel plungers (punches or platens) and is then ejected by upward movement of the bottom plunger after the top plunger is removed. The pressing cycle is shown in Figure 4.10.

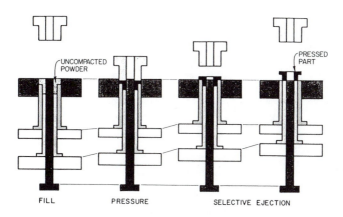

4.10. Details of dry pressing cycle for small parts. Note that the die incorporates a stationary central mandrel (to form the hole) and two concentric bottom punches that slide past one another as the cycle progresses. The actual pressing occurs in the second stage, while the last two stages involve ejecting the part (*courtesy AC Compacting Presses, Inc.*).

In the refractories industry, large toggle presses, such as that shown in Figure 4.11, form several bricks during each pressing cycle, using multiple-die cavities.

In forming many technical ceramics, small presses, such as shown in Figure 4.12(a), produce thousands of parts each day, mainly for electrical and electronic applications. Presses ranging in size from those capable of exerting 2 tons up to those capable of 100 or more tons are used to produce articles ranging from transistor heat sinks to grinding wheels (Fig. 4.12[b]).

Important considerations for successful dry pressing include

1. Parts to be pressed should have a low length-to-diameter ratio to ensure even compaction pressure throughout the die and uniform density throughout the part.
2. The particle size distribution of the ceramic powder should be adjusted according to the nature of the parts being pressed. A distribu-

4.11. Semidry pressing of refractory shapes (*courtesy Chisholm, Boyd & White Co.*):

(a) Toggle press.

(b) Close-up of multiple die during ejection of pressed brick, showing top punches.

(c) Exchangeable dies permit convenient changeover to different shapes.

4.12. Dry pressing:

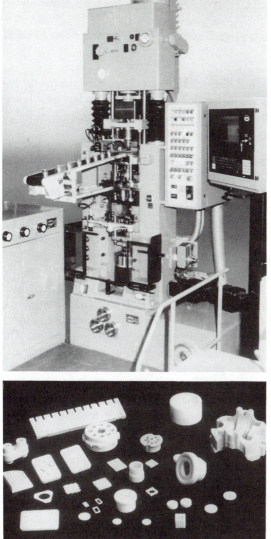

(a) Press used in dry pressing small ceramic parts (*courtesy Dorst America,* Inc.).

(b) Some shapes formed by dry pressing (*courtesy AC Compacting Presses,* Inc.).

tion of very fine particles is required for thin ceramic wafers. Blending of variously sized particles to accomplish efficient packing is important to the production of high density.

3. The binder composition should be adjusted to the pressing pressure and the parts being pressed. For example, wax binders are suitable for low pressures but harder, less tacky binders may be used at high pressures.

4. The pressing mix should be free-flowing and have no tendency to collect a static charge.
5. Sharp angles in the final pressed part should generally be avoided to prevent cracking.

Wax binders and stearate lubricants are particularly useful in dry pressing. As in other forming operations, a blend of binders and lubricants is usually more desirable than using a single additive.

Isostatic pressing

In the ideal dry forming method, pressure would be exerted uniformly on all surfaces of the powder mass and throughout the formed part. These conditions cannot be realized in conventional dry pressing in steel dies, but isostatic pressing offers a considerable improvement. In this process a dry or semidry granulated ceramic mix is placed in a pliable rubber or polymer mold (sometimes called the *tooling*), which is sealed, evacuated, and then squeezed uniformly by immersion in a high-pressure oil or water cylinder.

To form a hollow shape, a metal mandrel is usually placed in a rubber mold, and the granulated ceramic mix is poured between the mandrel and the walls of the pliable mold. Vibration is often used during mold filling to help pack the mix uniformly into the mold cavity. The mold is then sealed, evacuated, placed in the high-pressure cylinder, and gradually pressed to the desired pressure. The forming pressure is gradually reduced before the part is removed from the mold. The inner contours of the piece are formed by the mandrel and the outer contours are formed either by the inner contours of the rubber mold or by later machining before firing.

Large shapes are generally formed in water-filled presses (Fig. 4.13), some of which are capable of exerting pressures up to 50,000 pounds per square inch (psi) or more on the ceramic part. Small shapes, such as spark plug insulators, and dinnerware can also be rapidly formed, often with multimold presses.

Some considerations for successful isostatic pressing are

1. The tooling (molds and mandrels) should be carefully designed taking into consideration the flow properties of the ceramic mix, the pressing pressure and anticipated firing shrinkage, the amount the part will spring back from a pressing mandrel, and the desired configuration of the final fired part.
2. The binder system should be similar to that for dry pressing, but hard binders should generally be avoided.

4.13. Isostatic press for large refractory shapes. The heavy top closure for the high-pressure chamber is raised and swung away by the crane arrangement at the upper left (*courtesy National Forge Co.*).

3. Vibration and vacuum should be used during the mold filling operation for most parts.
4. The rate of pressure application and release must be controlled according to the nature of the part being pressed. (Generally, greater care is required for large parts than for small ones.)

LP gas such as propane has been used to help in the pressure release process for large shapes (replacing air, which is not very compressible). Use of such a gas requires certain safety precautions because the gas can settle in low areas at explosive concentrations.

Because of the large production volume of dinnerware shapes, the high machinery and die cost associated with dry isostatic pressing can be justified. The shape is formed by a membrane pressing the spray-dried granules (which usually contain a few percent moisture) against a steel plate (Figs. 4.14 and 1.3[c]). The properties of the membrane determine the quality of the product. For less critical products, the material is simply pressed between steel dies.

Explosive forming

Explosive forming consists of placing a granulated ceramic mix around a steel forming mandrel inside a pliable envelope (frequently a

4.14. Fully automatic production system for the manufacture of isostatic pressed flat tableware (*courtesy Dorst America, Inc.*).

thin metal foil), packing explosives around the outside of the envelope, and igniting the explosive (usually under water). This process has had limited use for ceramics but is occasionally used for the fabrication of refractory metals and carbides.

High-temperature forming

Some ceramic materials can be formed at high temperatures by techniques very similar to those utilized by the metal industries. These methods, with the exception of glass forming, which will be discussed in the next section, are rare in the ceramic industry at present, but will undoubtedly have an increased popularity in the future as improved techniques are developed.

Hot forging

Hot forging consists of heating previously dry pressed materials to high temperatures and then pressing them in cold steel dies. The formed objects are then returned to the furnace for annealing — the process of releasing the residual stresses caused by the additional pressing. This method gives improved density and strength over conventional cold pressing and has drawn some interest from the refractories industry.

Hot extrusion

Hot extrusion may be used with those ceramic materials such as MgO which have reasonable ductility at high temperatures. Special tooling and dies are required for this process, and so far it has had limited application. The process is similar to extrusion methods used to form metals.

Hot pressing

Ceramic powders may be pressed at high temperatures in heated dies made of graphite or other materials having good hot strength. The process combines the forming and firing operations. The process is slow but often results in a product superior in strength and density to products formed by cold dry pressing and subsequent firing. This process has fair popularity in the many specialty areas of ceramic manufacture.

Since graphite or refractory metal hot pressing dies are electrical conductors, they usually are heated by induction, although resistance heating is sometimes used. The high-frequency induction field is usually supplied from a motor-generator set through a water-cooled copper coil surrounding the die. Zirconium oxide grain is often used as thermal insulation between the die and the copper coil with the grain being held in place by a refractory pipe. Figure 4.15 shows the main components of a hot pressing operation.

Pressures of only a few thousand psi are needed to press at high temperatures. Thus the main costs in hot pressing are the electric power supply and labor for this rather slow process. There also is a substantial cost in the graphite dies which must often be discarded after pressing only one part.

Hot isostatic pressing

Gas pressure bonding or hot isostatic pressing (HIP), one of the most promising innovations in high-temperature forming methods, utilizes a combination of high-pressure inert gas and high temperature to densify ceramic parts. The method is similar in principle to regular isostatic pressing. Powdered ceramic material or a preformed part is first encapsulated in a pliable refractory metal foil container, often Mo. The container is then placed in a pressure vessel containing a high-temperature resistance furnace. The chamber is first purged with an inert gas and then evacuated so that all traces of air and moisture are removed.

At the beginning of the bonding cycle, helium or other inert gas is injected into the pressure vessel at the desired forming pressure. The furnace temperature is raised according to a desired bonding cycle and maintained at the final bonding temperature for a desired time. The

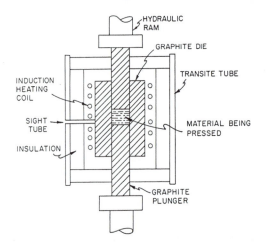

4.15 Hot pressing:

(a) Schematic of an induction-heated hot press.

(b) Exterior view of vacuum hot press (*courtesy Vacuum Industries*).

(c) Interior view of vacuum hot press (*courtesy Vacuum Industries*).

furnace is cooled, and pressure is vented according to a controlled program to prevent cracking of the pressed parts. The crushed metal foil container is then removed from the part and discarded. HIPing has also been successfully performed using a soft glass for encapsulation that is applied to a preformed part as a powder and then fused into a gas-tight envelope during heat-up in the HIP apparatus. HIP equipment is available to form materials at up to 15,000 psi and 3000°F (1650°C) (Fig. 4.16).

4.16. Hot isostatic press (HIP) (*photo by John T. Jones*).

Fusion casting

A high-temperature forming method of some importance in the refractories industry is fusion casting. Powdered refractory ceramic materials to be fusion cast are dry-blended and then completely melted by carbon-arc electrodes. The melt is poured into refractory molds of the proper shape and allowed to slow-cool, yielding an extremely dense and impervious coarsely cystalline block. During crystallization, a shrinkage "pipe" or hollow depression forms in the upper part of the block, but this is usually cut off with a diamond saw. Fusion-cast refractories are extremely resistant to attack by molten glass and are routinely used in the lining of glass melting tanks.

Thermal spraying

Ceramic coatings can be applied to substrates in a process analogous to spray painting by feeding ceramic rods or powders into a thermal spraying gun that melts the ceramic and propels the molten droplets at high velocity to the target. The resulting coatings can protect the substrate from corrosion, abrasion, or excessive temperature.

The basic thermal spray gun types are plasma, combustion-flame, and electric arc. The most versatile process utilizes the plasma gun, which is capable of melting any ceramic material and spraying at velocities of greater than 200 m/second. This results in good-quality coatings, potentially approaching theoretical density.

The best coatings on metal substrates are obtained by plasma spraying in a reduced-pressure chamber. By restricting the presence of oxygen, the metal/coating interface can be maintained at high temperature without oxidizing the metal, promoting diffusion at the interface and resulting in excellent bonding. The insides of tanks can often be robotically sprayed at reduced pressure, simulating the ordinary use of a reduced-pressure chamber.

LANXIDE ceramic composites

The advent of LANXIDE ceramic composite articles, which has been compared in significance with the development of glass-ceramics, was made possible by the invention of a process that grows ceramic composites matrices within a structure of preplaced reinforcement materials using oxidation reactions between molten metals and adjacent oxidants. The oxidized metal forms the ceramic matrix of the composite.

In the DIMOX directed metal oxidation version of this process (Fig. 4.17), molten metal alloys are oxidized to form the matrix by wicking of the metal through the growing reaction product layer toward the oxidant. The reinforcement preform can be fabricated by conventional processes. The dimensions of the preform do not change during the matrix formation process; however, a growth barrier should be used to cover the preform on surfaces exposed to the oxidant (usually air or nitrogen). Selection of growth alloys and filler materials is important, and special processing may be required. Some residual unoxidized metal is typically present in the final composite, although for some applications the metal is removed or specially treated. Examples of composites being made are listed in Table 4.1.

TABLE 4.1. **Composite systems made using directed metal oxidation process**

Matrix	Filler
Al_2O_3/Al	Al_2O_3, SiC, $BaTiO_3$, SiC fiber
AlN/Al	AlN, Al_2O_3, SiC, B_4C, TiB_2, SiC fiber
ZrN/Zr	ZrN, ZrB_2
TiN/Ti	TiN, TiB_2, Al_2O_3
Si_3N_4/Si	SiC, Si_3N_4, C fiber

4.17. LANXIDE process (*courtesy LANXIDE Corp.*):

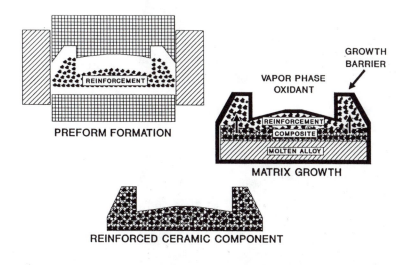

PREFORM FORMATION

MATRIX GROWTH

REINFORCED CERAMIC COMPONENT

(a) Schematic illustration of the growth to net shape of a composite using directed metal oxidation (DIMOX) through a preform of ceramic powder.

(b) Optical micrograph of an Al_2O_3/Al matrix grown through a SiC particulate reinforcement. Similar composite structures have been produced using ceramic fiber reinforcements.

(c) Growth of Al_2O_3 reinforced ceramic products in an air atmosphere furnace.

(d) Products are directly fabricated to final or near-final shape and size. Here, hydrocyclone liners undergo final sizing, requiring only minor grinding of top and bottom edges.

(e) Wear-resistant components fabricated of SiC particulate-reinforced Al_2O_3. The largest part shown was grown in one piece to a diameter of 686 mm (27 in.) and weighed 76.5 kg (169 lb).

Advantages of directed metal oxidation fabrication technology include net or near-net-shape capability, simple layup and processing, fast reaction times, and ability to engineer microstructures and properties. Examples of components made from LANXIDE composites are shown in Figure 4.17(e). Commercial use or evaluation has been made for such applications as slurry pumps, hydrocyclones, armor, gas turbine engines, rocket engines, piston engines, heat exchangers, and high-temperature furnaces.

Glass forming

A glass is produced when certain raw materials — usually including at least one silicate, borate, or phosphate mineral — are melted together to form a viscous liquid at high temperature. If this liquid solidifies on cooling without turning into a crystalline mass, it is called glass. As the cooling takes place, the viscosity of the melt rapidly increases until, at a temperature called the glass transition temperature (which is below the temperature that crystals normally start to form), the material becomes effectively rigid and is correctly called a glass. If the glass is reheated, it will gradually decrease in viscosity until it becomes soft enough to begin to flow. The ability to control the viscosity of a glass by careful temperature control is the key to all glass-forming operations. Certain glasses may be transformed into crystalline masses by further heat treatment after they have been formed by normal glass-forming techniques; these materials constitute the class of materials known as glass-ceramics.

Glass fibers

Two basic types of glass fibers are manufactured. Long continuous fibers are used for reinforcement in polymer matrix composites, for weaving into textiles, and as wave guides for fiber optic communications. Short fibers are used in thermal insulation.

Continuous fibers are drawn through small orifices in a heated platinum bushing, an operation requiring a bubble-free molten glass of very uniform composition. For many years the raw materials were premelted, refined, and formed into glass marbles. These marbles were then charged into the platinum bushing for remelting and forming. Recent developments in furnace design utilizing convection currents to stir and refine the glass have permitted the elimination of the costly marble-forming and remelting steps. The fibers are drawn through the bushing orifices onto high-speed spools.

Long fibers can be woven into cloth sheets or tapes which may be impregnated with a thermosetting resin. Rigid shapes of such fiberglass composites are built up by wrapping these tapes around a mandrel or by shaping the sheets around a form. Often strands of glass fibers are chopped, mixed with resin, and sprayed into a mold. In all cases, the formed article is gently heated to cure the resin.

Glass fibers intended for wave guides in optical communication are given a special coating during drawing. The difference in index of refraction between the glass fiber and the coating assures that light signals will travel down the fiber by total internal reflection at the fiber-coating interface. Truly remarkable levels of chemical purity, uniformity, and freedom from flaws are required in glass fibers intended for long-distance communications.

Short glass fibers for insulation can be made in several ways. Coarse fibers can be extruded from a bushing, chopped, and attenuated into fine fibers by a steam or burner blast. A more recent development uses a burner blast to attenuate fibers thrown out from a rapidly spinning centrifugal rotor. The short fibers can be felted into resin-bonded mats for use as thermal insulation.

Glass containers

Hollow shapes suitable for containers have been formed for many centuries by simply using air to blow a bubble in hot fluid glass. Using only hand tools and their lungs as a source of air, skillful artisans have long made intricate hollow shapes such as bottles, goblets, and complex chemical apparatus. Today, with the exception of a few very large low-production-volume shapes, most nonart glass containers are formed by machine (Figs. 4.18–4.20). Molten glass is fed to these machines either by cutting a flowing glass stream into individual *gobs* or by "gathering" onto a pipe using reduced pressure. The shaping is done by blowing the soft glass gob into a cold steel or bronze mold. A single glass-melting furnace can feed a number of forming machines. The rapid cooling of the molten glass during forming introduces residual stresses that would cause the container to fail in use. These stresses are removed by reheating the glass article below its softening point in an annealing *lehr* (furnace) where the stresses are allowed to relax. The annealing is followed by slow cooling.

4.18. Bottle-forming machine. Individual molten glass gobs fall from the melting furnace above into molds where they are blown to shape (*courtesy Owens-Illinois, Inc.*).

4.19. The blowing operation is performed in two steps. The gob is first blown into a rough shape called a "parison," which is then transferred to a final blowing mold. In the photo, hot parisons are being transferred to the final mold, while fully formed bottles leave the machine (*courtesy Owens-Illinois, Inc.*).

4.20. Various shapes can be formed by blowing. Here a turret chain machine swings a mold into place around the gob to form a large hollow shape (*courtesy Corning Glass Works*).

A second forming process for shallow ware or flatware of small size consists of pressing a softened gob of hot glass into a steel mold (Fig. 4.21). Items such as ashtrays, lens blanks, and automobile headlamp covers are made in this manner; again, annealing is necessary following the forming operation.

4.21. Shallow glass shapes can be formed by pressing hot glass gobs into molds (*courtesy Corning Glass Works*).

Flat glass

Glass for windows in buildings and automobiles is produced by allowing the molten glass to form a continuous sheet by pouring it out from the glass-melting furnace onto the surface of a shallow bath of molten tin that is contained within a hot enclosure at the end of the melting furnace. This is called the float process because the glass has a much lower density than the molten metal, which causes it to float on the surface of the metal bath. Because of the high temperature, the glass readily spreads out into a thin layer on the molten tin. The surface of the molten tin is very smooth, so it forms a smooth "bottom" on the molten glass layer. The top of the molten glass layer becomes fire-polished—that is, becomes smooth because of surface tension. As the glass moves away from the furnace along the tin bath, it gradually cools and solidifies to become a continuous sheet of glass of uniform thickness and extremely smooth surfaces. After passing through an annealing lehr to remove strains, the continuous sheet (see Fig. 1.2[a]) is cut into desired sizes by diamond scribing (Fig. 4.22) and is inspected and packaged for shipping. The combination of the melting furnace, tin bath enclosure, and annealing lehr, all in a line, requires a very long factory.

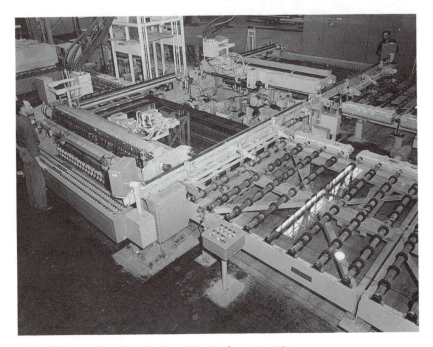

4.22. After annealing, flat glass is cut to size by automatic machines (*courtesy PPG Industries*).

Tempered glass and laminated glass

Glass products can be strengthened to four or five times their annealed strengths by rapid air-jet cooling from about 650°C (1202°F). This tempering process puts the surface of the glass into residual compression. In a situation analogous to prestressed concrete, the glass will not fail under load until tensile stresses are reached that equal the sum of this residual compressive stress plus the inherent tensile strength of the glass. On failure, tempered glass does not shatter in the normal way, but rather forms many small, rounded, harmless fragments. Tempered glass objects cannot be cut and must be completely formed before tempering. Glass can also be tempered by chemical surface treatments, which result in surface compressive stresses.

Curved laminated automobile windshields are shaped by heating matched sections of flat glass in stainless steel molds. The softened glass slumps and takes the mold configuration. After annealing and cooling, dry polymeric laminating film is sandwiched between the matched glass halves. The halves are pressed together under heat and the polymer is then bonded to the glass in an autoclave at 100°C (212°F) under moderate pressure. Such laminated glass does not shatter in the normal sense because the laminating material keeps all fragments and slivers in place.

Glass-ceramics

Glass-ceramics are formed as glasses using traditional glass-forming techniques, followed by special heat treatment to allow controlled crystallization of the glass to form a 90% to 98% crystallized ceramic. The crystallite size is from 0.1μm to 1μm. Several compositions can be used depending on the end use of the glass-ceramic. One of the most important compositions is based on the Li_2O-Al_2O_3-SiO_2 system.

Typically, the product is formed as a glass, cooled, then reheated to and held at an appropriate temperature where a very high concentration of minute crystal nuclei can form in from one to two hours. It is then heated to and held briefly at a higher temperature where the final crystallization and grain growth can occur. Finally, the product is cooled to room temperature.

A nucleating agent dissolved into the glass is required for the subsequent crystallization process. Zirconia or titania particles that come out of solution are often used. An interesting variation uses tiny metallic silver particles as the nucleating agent, with the silver particles being formed "photographically" by exposure of the glass to light. This allows photographic patterning of the crystallized portion of the glass for interesting decorative and etching applications.

5 Drying, firing, and

MANY CERAMIC FORMING PROCESSES require the intentional addition of water to the material mixture. Since the ware thus formed will be fired at high temperatures, which would turn this water to steam and literally explode the ware, the water must be removed in a controlled fashion prior to firing. Drying is the removal of mechanically combined water and sometimes also involves vaporization of organic additives such as plasticizers and binders. Ceramic products formed without the addition of water, including many electronic components, certain refractory shapes, and spark plug insulators, do not require a drying step prior to firing.

Firing (sometimes called burning) of ceramic ware is one of the most important steps in the overall production process. It is during firing that the "green" ware "matures"—that is, the final properties and the final usefulness of the ceramic product are developed. Detailed descriptions of the physical, chemical, and mineralogical changes that take place while the material is maturing are complex, and only a general description of these processes is given here.

A variety of furnaces (often called kilns) are utilized to carry out the firing of ceramic ware and the melting of glasses. A representative selection of these will be described, along with a brief discussion of combustion and furnace

finishing ceramic ware

atmospheres. The methods used to finish ceramic ware after firing, including grinding, glazing, and bonding to other materials, will also be described.

Drying

The removal of water from ceramic ware can take place only at the surface of the ware by evaporation. The water on the interior of the ware must travel to the surface by seeping or wicking through interconnected pores. Both processes — evaporation and seepage toward the surface — are accelerated by heating. Furthermore, the rate of evaporation of water from the surface is accelerated by low surrounding humidity and by rapid air movement across the surface.

Since the individual particles that make up the wet-formed ware are held apart by thin layers of water, the removal of this water will cause the particles to pull together, resulting in an overall decrease in the dimensions of the ware. This drying shrinkage is discussed in some detail in Chapter 8. The greater the amount of water originally added to aid in forming the ware, the greater will be the amount of drying shrinkage occurring when that water is removed. This shrinkage must be taken into account when the ware is originally formed.

If the rate of evaporation of water from the surface of a drying piece is greater than the rate at which water can seep (or diffuse) through the pores from the interior toward the surface, the air/water interface will move inward and the surface of the piece will dry faster than the interior. This very dangerous condition (sometimes called *case harden-ing*) causes the surface layers of the piece to shrink considerably while the interior remains virtually unchanged, and a network of tension cracks will almost certainly occur across the surface. A similar situation can develop in a piece having both thin cross-sections and thick cross-sections. Here the thin cross-sections may dry completely before the thicker sections, and the difference in shrinkage between sections would probably cause cracking to occur.

The solution to the case-hardening problem lies in inhibiting rapid evaporation of water from the surface of the piece while large amounts of water still remain in the interior. This can be accomplished by heating the piece in an enclosure where the relative humidity initially is kept high. The higher the humidity of the air surrounding the drying piece, the slower will be the evaporation rate from the surface of the piece. Combining gentle heating to encourage seepage toward the surface with high humidity to suppress evaporation results in a situation where the rates of these two processes are nearly equal and the piece dries uni-formly. As drying progresses, the temperature can gradually be raised and the humidity gradually lowered so that evaporation and seepage rates remain reasonably high. The final stage of drying will be done at very low humidity and at temperatures exceeding 212°F (100°C), the boiling point of water. This should result in a bone-dry piece that is free of shrinkage cracks. It may not be necessary to have ware absolutely bone dry prior to firing.

The problem of preventing shrinkage cracks due to nonuniform cross-sections cannot be overcome by this procedure. Instead, the thin-ner cross-sections are usually prevented from drying faster than the thicker sections by selectively wrapping the thin sections with moist cloths or spraying them with waxy coatings that suppress the rate of evaporation. This kind of drying can be done either at room temperature in the open or at warmer temperatures if the humidity is kept high. Whenever ceramic pieces of thick cross-section are to be dried and a controlled humidity enclosure is not available, the use of moist rags or surface coatings can be used to prevent formation of drying cracks due to surface shrinkage (case hardening).

Since air at normal drying temperatures holds only a limited quan-tity of water before reaching saturation (100% relative humidity), and since evaporation takes place only if the air in contact with a drying piece is at less than 100% humidity, fans are used to constantly circulate air over a drying piece to sweep away the saturated layer nearest the surface. If the rate of air circulation is increased without changing its humidity or temperature, the rate of evaporation will increase. Thus the

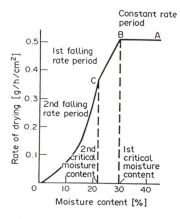

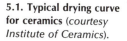

5.1. Typical drying curve for ceramics (*courtesy Institute of Ceramics*).

important variables to control in any drying operation are air temperature, relative humidity, and air flow rate.

A final important consideration in driers is the dew point of the air being introduced. The dew point for a moisture-containing gas such as air is the temperature at which it will start to condense out water. The higher the moisture content of air, the higher its dew point will be. If moist air encounters a solid that is at a temperature below its dew point, water will condense out of the air onto the solid. Thus very high-humidity air passing through a relatively cool portion of the drier system (such as a metal flue) may condense out water, resulting in serious corrosion problems. The dew point problem is also a serious one in the initial heating stage of ceramic firing.

Figure 5.1 illustrates a typical drying curve for ceramics. In this graph, the drying process starts at point A and proceeds toward the left (lower moisture contents). At first the rate of evaporation is independent of moisture content (*A–B*). This is the *constant rate period*. The *first falling rate period* (*B–C*) often shows a linear dependence between drying rate and moisture content. The moisture contents at the end of these periods are called, respectively, the *first critical moisture content* (point *B*) and the *second critical moisture content* (point *C*). In a typical situation where heat is transferred to the ceramic by air flowing over the surface (convection), the surface acts as a free water surface during the constant rate period, and the evaporation rate is not a function of the condition of the material below the surface. During the first falling rate period, the rate of evaporation can still be influenced by the rate of air flowing over the surface, but while liquid water is still moving from the interior of the ceramic, the surface no longer behaves as a free water surface. A *second falling rate period* (starting at point *C,* not always distinct) has a curvilinear relationship between drying rate and moisture content. During this period, the rate of evaporation from the surface of the ceramic is dependent on the diffusion rate of water from the interior of the ceramic and this slows as the diffusion pathway lengthens as drying progresses.

The amount of shrinkage occurring during drying can be quite different for nonclay and clay-bearing ceramics. In clay-bearing ceramics, the particles are separated by water layers before the drying process begins but shrink together during drying. Nonclay ceramic particles are often touching at points even before the drying process begins, and so drying shrinkage is usually less.

Ceramic ware of thin cross-section and low water content can be dried by simply placing in a warm area with good air circulation. Most ware that requires drying, however, must be processed in special drier enclosures where the temperature, humidity, and air flow can be controlled. The source of heat for these driers is generally a set of gas or oil burners, although often a considerable amount of waste heat from kilns is also transferred to the drier. Some driers are heated by infrared lamps, but electrical-resistance heating is too expensive for anything but the smallest production and laboratory driers. Both microwave drying and vacuum-assisted drying are receiving increasing interest today. For ease of handling, the ware is usually stacked onto cars that can be wheeled in and out of the drier on rails.

Periodic (batch) driers are necessary when a great variety of articles are being produced and fired in periodic kilns. These driers consist of a single chamber which is filled with ware and then cycled through a specific time-temperature-humidity schedule to achieve complete drying. The drier is then at least partially cooled and the dried ware is removed. When the drier is empty, a new charge of wet ware is rolled in, and the cycle is repeated.

Tunnel driers (Fig. 5.2) are continuous driers that are best suited to a continuous firing operation. They consist of a long chamber through which the ware is slowly pushed. Various zones of constant temperature and humidity are maintained along the length of the tunnel. The formed ware is usually placed directly on refractory-topped kiln cars which can then be passed through the drier and directly into a tunnel kiln without further handling of the ware.

It is sometimes necessary to dry granular ceramic materials in the unconsolidated condition. For drying moist agglomerated materials such as shredded filter cakes or the products of various wet separation processes, tray driers may be used in which the material is spread on trays that move progressively through a drying chamber. Such materials can also be dried in a rotary drier (Fig. 5.3), which consists of a long cylinder tilted slightly from the horizontal and slowly rotated about its axis. Hot dry air is blown into the lower end and exhausts from the higher end. The moist material is fed into the higher end and, because of the action of the rotating, tilted cylinder walls, slowly tumbles its way to the lower end where it is discharged dry.

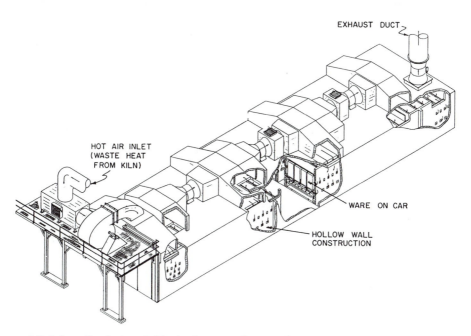

5.2. Schematic of a tunnel drier having several zones of controlled humidity and temperature along the length (*courtesy Swindell-Dressler Co.*).

5.3. Rotary drier. These driers are installed so that the "right" end is slightly higher than the "left" burner end (*courtesy Davenport Machine & Foundry Co.*).

If a ceramic slurry or slip is to be dried, filtering and conventional drying can be eliminated by spray-drying as discussed in Chapter 3.

Firing

The maturing of a ceramic body may take place in one of two ways: if formation of a large amount of glass by partial melting occurs during firing, the maturing process is called *vitrification;* if very little or even no liquid is present during firing, the maturing process is called *sintering.* In either case, the end results of firing are the same—that is, the reduction or near elimination of pores, accompanied by shrinkage and increased density, and a bonding together of individual crystalline grains or mineral constituents into a strong, hard mass.

The great majority of ceramics, including all whitewares, structural clay products, and fire clay refractories, undergo vitrification during firing. The terms *unvitrified, semivitrified,* and *vitrified* are descriptive of the amounts of porosity remaining in the final product. (See ASTM C 242–90a.) Most of the vitrifiable ceramic wares contain clays or talc along with other silicates. As the temperature of the body is raised, carbonaceous impurities burn out, chemical water is evolved, and carbonates and sulfates begin to decompose. All of these processes produce gases that must escape from the ware by passing to the surface through interconnected pores. On further heating, some of the minerals begin to break down into new forms, and the fluxes present react with the decomposing minerals to form viscous liquid silicates or glasses. (If this liquid formation proceeds far enough to begin blocking pores before all gases are vented, the body will swell up in an undesirable process called *bloating.*) More liquid forms as the temperature is raised, and it begins to pull the unmelted grains together by surface tension forces, causing shrinkage and an increase in bulk density. If the glass-forming process is allowed to go too far (either to too high a temperature or for too long a "soak" time at the temperature), so much of the mass will become liquid that it will no longer support its own weight. If this occurs, the mass will deform or slump, and the article will become worthless. Slumping can sometimes be eliminated by supporting the piece with special refractory shapes called *kiln furniture.*

When the proper degree of maturity—the proper density and amount of remaining porosity—has been achieved, the article is cooled. The cooling causes the liquid glass to become rigid and form a strong bond between the remaining crystalline grains. Figure 5.4 shows the microstructure of ceramics that have undergone vitrification.

The role of the potter's flint (silica) grains in a vitrifying porcelain ceramic is a critical one. Silica undergoes several changes in crystal structure during heating and cooling. These changes, called polymorphic transformations, result in an abrupt change in volume with a consequent tendency to cause cracking of the grains; such cracks can propagate through the ware because the grains are tightly bonded into the structure. Ordinary quartz transforms abruptly into a form called high quartz when heated to 573°C (1063°F). The sudden large volume change does not cause serious strains to develop in the unmatured ceramic mass. When liquid is present, the high quartz can very slowly convert into

5.4. Microstructure of vitrified ceramics:

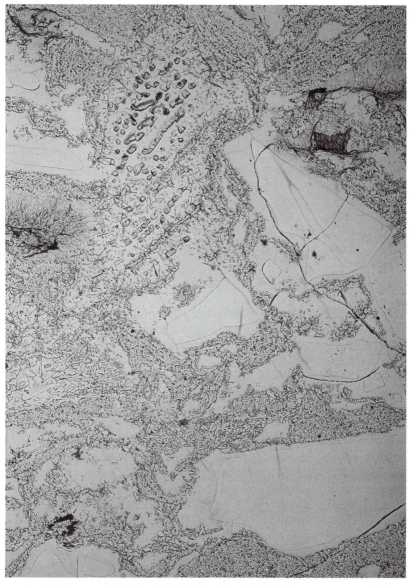

(a) Porcelain, showing large quartz grains surrounded by glass.
Very fine needlelike crystals are mullite. Major areas of mottled
appearance are remnants of the decomposed clay matrix.
Cracks in quartz grains are due to polymorphic transformation
on cooling (*courtesy American Standard*) (approximately 2200 ×).

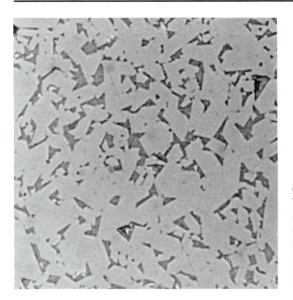

(b) High alumina body. The light gray areas are alumina grains; the darker gray areas show glassy bonding phase (*courtesy V.E. Wolkadoff and R.E. Weaver, Coors Porcelain Co.*) (400×).

other crystalline forms on further heating, specifically, to a form called tridymite above 867°C (1593°F), and to a form called cristobalite above 1470°C (2678°F). Some cristobalite is often formed in ceramics that have been fired to high temperatures, but much of the silica grains remain as unconverted high quartz. Since large silica grains remain relatively undissolved even at maturity, the probability that they will go back through the polymorphic transformations during cooling must be considered. When cooled below approximately 600°C (1112°F), high quartz grains will transform abruptly back into low (ordinary) quartz with a sudden volume change. In the relatively dense matured body that is fairly rigid at this low temperature, this abrupt volume change in the quartz grains on cooling can cause serious local strains to develop and may even cause cracking (*dunting*) of the ware. For this reason, cooling through the 600–500°C (1112–932°F) temperature range must be done very slowly. Finer grinding of the flint is sometimes helpful in overcoming this dunting problem. In fast-fire ceramics, alumina is often substituted for some or all of the potter's flint to avoid such cooling problems, because alumina does not undergo polymorphic transformations on heating and cooling.

Ceramics consisting of pure oxides such as Al_2O_3, MgO, $BaTiO_3$, and the ferrites, which do not contain any glass-forming constituent, cannot undergo vitrification. Instead, when an article consisting of a compacted mass of these oxide grains is heated, atoms move one at a time from points of contact between grains to pore walls, resulting in the bridging together of the individual grains into a coherent mass. This process, called sintering, may actually be accomplished by several different atomic mechanisms. Very high temperatures are usually necessary to carry out the densification at a reasonable rate. The rate of densification is usually found to be increased if the initial grain size is decreased. Often, small additions of other oxides can greatly increase the sintering rate of a pure material by forming small amounts of liquids at grain contact points. When most of the porosity has been removed, a gradual general coarsening of the grain size (grain growth) is usually observed. Figure 5.5 shows the microstructure of a sintered ceramic.

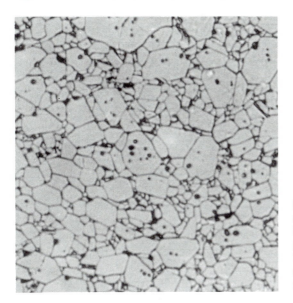

5.5. Microstructure of a sintered alumina ceramic. Small dark areas are porosity (*courtesy V.E. Wolkadoff and R.E. Weaver, Coors Porcelain Co.) (400×).*

Furnace atmosphere

The composition of the atmosphere in which a ceramic is being fired has a strong influence on the results of the operation, regardless of whether the ceramic matures by vitrification or by sintering. The oxygen content of the atmosphere inside the furnace is particularly important. If the atmosphere has sufficient oxygen to allow the ceramic to absorb some, it is said to be an oxidizing atmosphere; if the atmosphere tends to rob oxygen from the ceramic, it is said to be a reducing atmosphere. When the ceramic can absorb oxygen from the atmosphere, it can burn out carbon and convert all unstable compounds present to oxides. If reducing conditions occur, variable-valence ions in the ceramic will tend to change to their least positive valence, resulting in changes in optical (especially color), electrical, and other characteristics. The potency of fluxes can also be influenced by the oxygen content of the furnace.

In furnaces heated by combustion, the furnace atmosphere will contain a mixture of CO_2, N_2, CO, O_2, H_2O, and usually some SO_2. The SO_2 is often harmful to ceramics, especially if present in large quantities. When this is the case, common practice is to place a *muffle* in the furnace. This is a protective inner enclosure in which the ware can be placed so that the combustion gases flowing around the outside of the muffle never actually contact the ware. The muffle also protects the ware from direct flame impingement which can cause differential shrinkage during firing. If a muffle is not available, or if extra protection is required, *saggers* — individual closed refractory containers — are utilized to protect the firing ware.

Ceramic kilns

The furnace in which firing takes place, commonly called a kiln, can be classified as either a periodic (intermittent or batch) kiln or a tunnel (continuous) kiln, depending upon its construction and mode of operation. A periodic kiln is one in which the entire furnace is heated and cooled in accordance with the particular firing schedule used for the ware; a tunnel kiln, on the other hand, maintains certain temperature zones continuously, and the ware is moved from one zone to another to accomplish the required time-temperature cycle. The periodic kiln is the more flexible type since its time-temperature cycles can be tailored to a wide variety of different ceramic products. The tunnel kiln is the more economical of labor and fuel (although fiber linings and kiln car insulation have made periodics quite efficient), but is relatively inflexible, being limited to firing long runs of one kind of product. Many plants employ both kinds of kilns, the periodics being used for special products, and the tunnels for standard high-volume product lines. Initial investment for a tunnel kiln is high, and more sophisticated control systems are usually required than for a periodic kiln.

Periodic kilns usually consist of a single large refractory-lined, sealed chamber having burner ports and flues for carrying away combustion products. Unless it is constructed with a muffle, the kiln heats the ware by passing the hot combustion gases through the loosely stacked ware. In a downdraft kiln (also called a beehive kiln if round in shape) (Fig. 5.6) the flat refractory floor is perforated with many openings

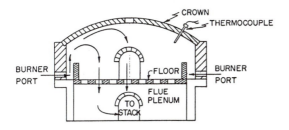

5.6. Side schematic view of a round downdraft kiln. These kilns are gradually being phased out of use.

leading to an underground flue system. The ware is loosely stacked on this floor so that vertical "chimneys" exist between pieces. The hot gases from the burners enter through the sidewalls, sweep up around the inside of the curved crown (roof), then pass down through the ware and into the flue system.

Care must be taken during early stages of heating that high dew point combustion gases do not begin to condense water onto the ware in the cooler, bottom part of the furnace. A less common periodic kiln design is the updraft kiln where the hot gases from a combustion plenum below the floor are distributed through a refractory grating and pass up through the ware to exhaust through the crown. Both types of kiln are set (loaded) and drawn (unloaded) by hand, resulting in considerable labor cost. They cannot be drawn until the kiln is cool enough to permit people to work inside, which means that very long cooling periods are necessary during which the kiln is unproductive. A total cycle time of several weeks is common for large, round, downdraft periodic kilns.

The shuttle kiln is a popular periodic type in which the ware is loaded onto a car that is rolled into the chamber for firing (Fig. 5.7). After firing, the shuttle kiln door is opened and the kiln car can be rolled out as soon as the kiln refractories and the ware have cooled enough to withstand the thermal shock involved in the process. In this way, the kiln need not be completely cooled and little time is lost between cycles.

Elevator kilns (Fig. 5.8) are also popular fast turnaround periodic kilns. The furnace shell (walls and crown) can be elevated (bell top kiln), while the hearth (bottom) remains on ground level. After the ware is placed on the hearth (often on a car rolled into place), the rest of the furnace is lowered around it. After firing, the furnace shell can be raised before the ware is completely cooled, shortening the long unproductive cooling period. In some elevator kiln designs the furnace shell is fixed above the plant floor, and the hearth is raised into place for firing.

5.7. Shuttles kilns (*courtesy Bickley, Inc.*):

(a) Large shuttle kilns each holding several kiln cars.

(b) Large, porcelain, electrical insulators fired in a shuttle kiln.

5.8. Elevator kilns (*courtesy Bickley, Inc.*):

(a) Bell-top production kiln with shell raised.

(b) Small elevator hearth kiln with hearth lowered.

119

The tunnel or continuous kiln (Fig. 5.9) consists of a refractory chamber, often several hundred feet in length, through which ware is slowly moved to accomplish gradual heating and slow cooling. The tunnel is always kept filled with ware along its entire length. The highest temperature zone, or firing zone, is generally near the middle of the tunnel length; this is where most of the burners are located. The first

5.9. Tunnel kilns:

(a) Kiln for firing brick showing loaded kiln car (*courtesy Bickley, Inc.*).

(b) Kiln for firing dinnerware (*courtesy Bickley, Inc.*).

section of the tunnel where the green ware enters the furnace is called the preheat zone. A few burners are located in this zone, with most of them concentrated near the beginning of the firing zone. The final section of the tunnel is called the cooling zone. (Actually, modern tunnel kilns and roller hearth kilns have a number of controlled subzones within each of these three major zones.)

(c) High-temperature roller hearth kiln (*courtesy Bickley, Inc.*).

(d) A display module of a roller hearth kiln (*photo by John T. Jones*).

Cooling air is forced into the cooling zone end of the kiln by large fans. This cool air travels over and through the cooling ware, picking up heat from it and accelerating the cooling process. The hot air resulting from this process blows into the firing zone, where it improves combustion efficiency and also assures an oxidizing atmosphere. Hot combustion products are blown from the firing zone into the preheat zone, where they lose heat to help bring the unfired ware up to temperature. Finally these cooled combustion products are exhausted through the crown of the preheat zone before they encounter ware that is cool enough to cause condensation of moisture. (See dew point discussion on page 109.) The resulting pattern of flow in the tunnel is as follows: the ware moves through the tunnel in one direction, picking up heat in the preheat and firing zones and giving it up in the cooling zone; gases move through the tunnel in the opposite direction, picking up heat from the ware in the cooling and firing zones, and giving up heat in the preheat zone. Very high thermal efficiency is the result of this complicated operation. Often outside air is blown through a double wall and double crown in the cooling zone to pick up additional waste heat for use in driers.

The movement of ware through large tunnel kilns is accomplished by using refractory-topped cars riding on rails. The cars are generally pushed into the preheat end on a fixed schedule. Since the tunnel is kept completely filled with cars, pushing one into the preheat end forces one out the cooling end—at the same time every car in the kiln moves one carlength forward. Sometimes a continuous refractory metal belt or a set of rollers is used in place of kiln cars to carry ware through a kiln. Roller hearth tunnels permit a very fast firing cycle and are more flexible than conventional tunnel kilns. The positively driven refractory rollers may be metal, mullite, or silicon-impregnated silicon carbide, depending on the temperature requirements of the kiln. The ware is usually flat, as with tile, or is placed on a flat refractory plate (batt) which moves forward through the kiln on these rollers. Roller kilns are used now in many sectors of the ceramic industry for firing a diversity of products from miniature electronic circuitry components to large sanitary ware items.

Unconsolidated ceramic materials can be calcined in rotary kilns, which are refractory-lined, rotating, cylindrical steel tubes (Fig. 5.10 and 1.7). Rotary kilns may be as much as 500 feet long and 12 feet in diameter. The tube is slightly inclined and rotates slowly on its axis. The burners are located at the lower end of the tube, and the material to be fired (in the form of a slurry, briquets, or loose powders) is fed into the elevated end. The rotating action tumbles the material down through hotter and hotter zones until the final desired temperature is reached just before discharge at the low end. The heat drives off any water or gases combined in the raw material and greatly densifies the individual ceramic particles. Many ceramic materials are calcined on a definite heating schedule, which preserves the reactivity of the material. This is particularly important in the production of reactive aluminum oxides and cements. Other materials, such as MgO, are "dead-burned" to minimum reactivity for use as aggregate in refractories and other ceramic products.

5.10. Rotary kilns (*courtesy Alcoa*).

Glass melting

The melting together of raw materials to form glasses is usually accomplished in large, shallow, pool-type furnaces called glass tanks. The bottom part of the tank is constructed of special refractory blocks and holds a pool of molten glass several feet deep. Above the level of the glass surface are short walls containing burner ports topped by an arched crown. Heating takes place primarily by radiation from the hot crown down to the glass. Large refractory-filled compartments, called checker chambers or regenerators, are placed at one end or along and below either side of the tank. The purpose of these regenerators is to capture some of the heat in the exhaust gases and use it to preheat the air used in combustion (Fig. 5.11).

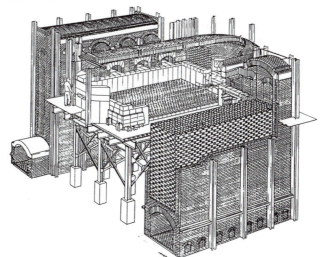

5.11. Cutaway view of a glass tank showing checker-filled regenerator chambers on either side. Raw materials are charged at left, melted, and pass through the throat in the bridge wall and into the refining zone at the right for conditioning prior to forming (*drawing courtesy Harbison-Walker Refractories Division of Dresser Industries, Inc.*).

A refractory wall, called the bridge wall, stretches across the width of some tanks and divides them into two distinct sections. In the larger melter section, raw materials are continuously fed in and melted to produce a bubble-filled viscous liquid. In the shorter refining section, the liquid glass is allowed to free itself of bubbles (to fine) and is conditioned to the proper viscosity for forming. In some tanks, the glass flows from the melter section into the refining section through a hole in the bridge wall. This hole, the throat, is submerged below the glass surface so that unmelted batch material floating on the surface in the melter does not find its way into the refining section. Other tanks use an anchored "floater block" of refractory to retain the unmelted batch. The steady flow of glass from the melter into the refining section occurs because glass is continuously being removed from the refining end for forming.

It is common practice to augment combustion heating of the tank with some electrical heating. This procedure, called boosting, uses the glass itself as a resistor and passes large electric currents through the glass pool between refractory metal electrodes immersed in the melt.

Specialty glasses and optical glasses are not normally melted in tanks. Instead, large refractory containers called pots are used for both the melting and fining of such glasses. The pots are placed inside a large furnace for heating.

Combustion

The source of heat for most industrial processes, including the drying and firing of ceramics, is the direct combustion of a fuel with air. The fuels most used in the ceramic industries are natural gas, fuel oil, liquid petroleum (LP) gases, producer gas, occasionally powdered coal, and rarely lump coal. All these fuels are made up of various combinations of the combustible chemical elements carbon and hydrogen with usually a very small amount of undesirable sulfur also present. When these fuels are burned by combining with oxygen from the air, they produce various gaseous combustion products plus heat. If carbon is burned completely, it forms CO_2; if combustion of carbon is not complete, a mixture of CO_2 and CO results. When hydrogen is burned, it forms water vapor (H_2O). Sulfur burns to produce SO_2. The properties of common fuels used in the ceramic industry are given in Table 5.1.

In order to promote more complete combustion and to generate large quantities of hot gases, air in excess of the quantity theoretically necessary for reaction is often used. The combustion products from a sulfur-free fuel burned with excess air usually consist of a mixture of the gases CO_2, CO, H_2O, O_2, and N_2. The N_2 comes from the air used to

TABLE 5.1. Properties of typical fuels

Liquid Fuels	S (%)	C (%)	H (%)	Ash (%)	Density (lb/gal)	Heating Value (BTU/gal)	Air Required (SCF/gal)
No. 2 fuel oil	0.10	86.8	13.0	NIL[a]	7.18	138,000	1381
No. 4 fuel oil	0.30	87.6	12.0	0.10	7.60	143,000	1437
No. 6 fuel oil	0.50	88.2	11.0	0.30	8.15	150,000	1513
Solid Fuels						(BTU/ft³)	(SCF/ft³)
Bituminous coal	1.78	75.3	5.0	8.00		13,000	136
Gaseous Fuels					Specific Gravity	(BTU/ft³)	(SCF/ft³)
Propane	0.00	82.0	18.0	NIL[a]	1.55	2500	23.90
C₃H₈ (97.3%)							
C₂H₆ (2.2%)							
C₄H₁₀ (0.5%)							
Natural gas	<1.0 VAR[b]	76.1	23.0	0.00	0.60	1000	9.56
CH₄ (90%)							
C₂H₆ (5%)							

[a]NIL = negligible.
[b]VAR = variable.

supply the oxygen for combustion. The relative amounts of these gases depend upon the amount of excess air and the overall combustion efficiency of the operation. It is common practice to monitor the chemical composition of the combustion products to optimize the operation. Many instruments are available to automatically monitor the furnace atmosphere for one or two of the more important gases such as excess oxygen and unburned combustibles.

Natural gas is the most convenient fuel to use. In order to make it more economical, it is sometimes purchased at a reduced rate on an interruptible basis. During times of limited gas supplies, especially in winter, the user's gas supply will be interrupted by the commercial gas supplier, and a standby fuel system will be needed. Such standby systems often employ fuel oil, although LP gas is also popular for such uses. Burners are available that burn both natural gas and the standby fuel, making quick changeover possible.

Finishing

Many ceramic products can be taken directly from the kiln, inspected, and shipped to the customer. Certain other products, however, require additional processing to meet customer specifications. These postfiring processes are grouped under the general category of finishing operations and may include grinding to meet critical size or surface finish requirements and application of adherent coatings for protection, decoration, or other special needs. Technical ceramics are carefully examined for flaws before shipment to the consumer. Cracks and pits can sometimes be detected visually, as shown in Figure 5.12, or by dye pene-

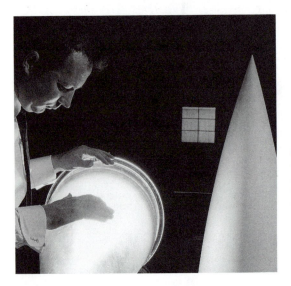

5.12. Inspection of a radome for flaws following precision grinding to contour (*courtesy Corning Glass Works*).

tration tests. X-ray and ultrasonic techniques may also be used to detect flaws.

Grinding

Most ceramic products can be fired to meet customer specifications on dimensions without further processing. By calculating the proper oversized dimensions of the formed part from dimensions specified by the customer, and by knowing the shrinkage of the ceramic associated with drying and with firing, a manufacturer can usually fire to a given size and shape specification. Frequently, special firing techniques are required to insure meeting a specification. The ceramic parts may have to be supported by refractory kiln furniture to prevent warping or slumping during firing. Solid refractory setters or setters of the same composition as the parts being fired may be placed under certain parts to help control shrinkage and warpage. A layer of refractory grain such as calcined clay or fused alumina may be required to keep parts from reacting with or sticking to setters or refractories during firing. Tubes and rods are frequently hung by collars into refractory saggers so that gravity tends to keep them straight during firing.

If the manufacturer cannot meet the dimensional and surface finish specifications of the customer by special firing techniques, then the ceramics must be ground and (or) polished after firing. Diamond tooling is usually required to grind hard materials such as alumina ceramics, but silicon carbide, alumina, or other abrasives may be used for softer materials. Disks may be lapped to the proper flatness, thickness, and surface finish (Fig. 5.13). Centerless grinders are often used to grind the outer

5.13. Lap polisher for flat ceramics and substrates (*courtesy Lap Master, Inc.*).

diameters of cylinders and rods. Small parts as well as the diameters of disks or the ends of cylinders may be ground on tool post grinders.

Very close size tolerances are commonly required on technical ceramic parts. The dimensions of these parts are frequently measured under carefully controlled humidity and temperature conditions with specialized measuring devices. Special surface finishes are sometimes required on ceramic parts. Special grinding and polishing techniques are always required under these circumstances.

Glazing

Many ceramics are glazed after firing. A *glaze* is a special glass coating designed to be melted onto the surface of a ceramic body and to adhere to that surface during cooling. Glazes are used primarily to seal the surface of a porous ceramic to prevent absorption of water or other substances. The resulting smooth impermeable surface is also attractive and easy to clean. In high-tension insulators, a glaze ensures maintenance of good electrical properties even in the rain. Special colors and textures can be developed within the glaze to provide decoration and sales appeal.

The major constituent of a glaze is generally finely ground silica, with the addition of constituents rich in alkalies such as sodium and potassium oxides to lower the melting point of the glaze and alkaline earths such as calcium oxide to impart chemical durability. Lead oxide and boric oxide are also frequent glaze constituents. The overall composition is always adjusted in order to control the thermal expansion of the glaze, which must be equal to or slightly less than that of the underlying body. Additives are used to color the glaze or make it opaque. Techniques for formulating a glaze of known composition are discussed in Chapter 9.

Most glazes are prepared by wet-grinding together various raw materials along with a specially prepared commercial *frit*. The frit is a glass containing all originally soluble materials, all coloring oxides, and those materials which are toxic in uncombined form. Frits are produced commercially by melting and quenching a glass made up of the required chemical constituents. Quenching may be accomplished by pouring the molten frit directly into water or by roll-quenching in which the frit is quenched between water-cooled steel rolls. After quenching, the frit is ground, dried, bagged, and shipped to customers for use in formulating glazes or enamels.

To make the glaze, the frit is placed in a ball mill along with clays and other insoluble materials and milled to a definite particle size with the proper amount of water. Binders such as dextrine or gum arabic may be added to aid in application. The milled glaze slip may be applied to the ceramic by spraying or dipping, and patterns can be added by printing (Fig. 5.14). Spraying is used in many automated processes. The powdered glaze dries rapidly on the ceramic surface, and special drying is usually not required before firing.

5.14. Glaze application (*courtesy Malking Ltd.*):

(a) Straight-line spray glazing machine for dinnerware. The operator loads ware onto rotating supports, which carry it through the spray booth at the right.

(b) Multistage pad printer. The pad picks up the image from an engraved plate to which colored glaze slip has been applied. The image is then transferred to the dinner plate from the pad. Four colors are standard.

Glazes can be applied to green ceramics and to completely vitrified ceramics; however, ware is generally *bisque-fired* before glazing. Bisque firing is a low-temperature firing that removes the volatiles from the ware and accomplishes part or all of the firing shrinkage, thus assuring better success in the glazing operation. The body and glaze are then finish-fired together (*glost-fired*). Electrical ceramic ware is frequently completely matured and then glazed at a lower temperature than the maturing temperature of the body. Low-cost items such as pottery can frequently be sprayed in the green state and the body and glaze can be matured together. (In the *china process* the ceramic body is first fired to maturity; then the glaze is applied and matured by firing at a lower temperature. In the *porcelain process,* the bisque-fired body and glaze are matured together in the final glost-firing operation.) It is usually necessary to warm previously vitrified ware before the glaze slip is applied or it will not adhere properly.

Various glaze defects can occur in the different stages of production. *Crawling* (uneven coverage) occurs when the glaze slip does not satisfactorily wet the body. This can often be corrected by changing or increasing the amount of organic binder in the glaze slip. *Crazing* (fine network of cracks) occurs if a matured glaze has a higher coefficient of thermal expansion than the body. *Shivering* (shearing away of the coating in spots) occurs when the matured glaze has too low a coefficient of thermal expansion compared to the body. Pitting can be traced to volatiles in the glaze or body. Descriptions of these and other defects and methods for correcting them are given in classic books by Singer and Singer (1964) and Parmelee (1951).

Metallizing

Ceramic parts sometimes need to be bonded directly to metal parts, a process that is especially common in the electronics industry (Fig. 5.15). The process consists of first metallizing the surface of the ceramic and then soldering or brazing the hardware to the metallized ceramic surface. In one process a mixture of molybdenum and manganese metal powders is ground in organic solvents and binders and painted or sprayed onto the ceramic surface. The coating is then fired onto the ceramic in a furnace containing a cracked ammonia reducing atmosphere (hydrogen plus nitrogen) at a temperature sufficient to form an adherent cermet layer. Metal parts with expansion coefficients near that of the ceramic may then be attached to the metallized (cermet) area using solder or brazing alloys of metals such as lead, copper or silver. The metallized areas are often plated with nickel, silver, or gold before soldering or brazing to improve wetting by the molten solder alloy.

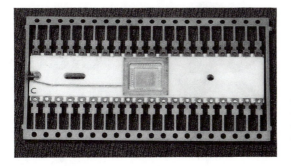

5.15. Example of a metal, electrical lead frame, which is soldered to metallized pads on a ceramic part (*courtesy Coors Porcelain* Co.).

Glass-to-ceramic seals

Many glasses have been developed to solder ceramics to other glasses. Techniques of application of the glass solder to the joint vary with the products being produced. Sometimes solder glass rings can be used to form the joint, and solder glass powders are used in other applications. The thermal expansion coefficient of the solder glass must be compatible with both the ceramic and the glass being bonded. The seal is made by heating the glass and ceramic parts together until they become hot enough to soften the solder glass. Glass-to-ceramic seals are extremely common in a wide range of products from automobile spark plugs to cathode ray tubes.

6 Selected

CERAMIC PRODUCTS UTILIZED for sophisticated technical applications today are so diverse as to almost defy classification. Often these materials are individually tailored to serve in only one specific application in order to optimize the particular set of properties required. Many of these modern technical ceramics exhibit properties that were never dreamed of by ceramists of the past. In general, the most sophisticated of these products are manufactured from very high purity raw materials (many of which are synthetic, and therefore quite expensive), often using novel forming techniques and special firing treatments, and frequently requiring extensive finishing and testing before being placed in use. To distinguish these sophisticated technical ceramics from the more traditional ceramics, the terms *advanced ceramics, fine ceramics,* and simply, *new ceramics* have been coined.

A partial listing of advanced ceramics would include high-temperature structural ceramics, ceramic cutting tools, bioceramic implant materials, space shuttle tiles (and other thermal barrier aerospace materials), ceramic armor, special aerospace window materials, optical fibers, very high permittivity capacitor dielectrics, sealing materials, multilayer ceramic substrates and electronic packages, piezoelectric transducer materials, electro-optic materials, "high-temperature" ceramic superconductor materials, ceramic nuclear fuels, a wide variety of ceramic coatings, solid electrolytes

applications for advanced ceramics

for high-temperature batteries and fuel cells, nonmetallic magnetic materials, thermistor and varistor materials, and so on. A relative newcomer to the advanced ceramics category of materials are the ceramic-matrix composites (also called CMCs). The use of ceramic reinforcements in polymer-matrix and metal-matrix composites has become fairly common; however, the formation of a composite that uses a ceramic material as the matrix (most often with ceramic fibers, whiskers, or platelets as the reinforcement) is a promising new area of development. A recent report by the U.S. Department of Defense Ceramics Information Analysis Center (CIAC) has predicted that the U.S. market for ceramic-matrix composites will grow at an annual rate of almost 15% to reach $500 million by the year 2000.

Whereas traditional ceramics are usually developed in industrial research and development laboratories, a wide variety of advanced ceramic materials have resulted from fundamental work performed in university and government research laboratories. These developments have occurred worldwide, with Japan being a leader in many instances. A measure of the importance to modern technology of advanced ceramics is indicated by the existence of such organizations as the United States Advanced Ceramic Association and the Electronics, Nuclear, Engineering Ceramics, and Glass and Optical Materials Divisions of the American Ceramic Society. The worldwide market value for advanced ceramics utilized in electronics, automobiles, integrated optics, advanced energy systems, bioceramics, and aerospace industries was approximately \$1.9 billion in 1985 and has been predicted to be about \$18.8 billion by 2000.

There are several keys to the successful application of advanced ceramics, in addition to their spectrum of useful properties. One is the development of new levels of understanding of the underlying science of the processing and fabrication of such materials that allows products to be made to extremely demanding size tolerances with a high degree of reproducibility in characteristics. A second key is the recognition that different design criteria are needed for hard, brittle materials than can be used for ductile materials. It has also been crucial to develop testing methods that yield "true" values for properties that can be used in critical design calculations, rather than the cruder testing methods which are suitable for making quality control judgements. (The property measurement techniques required for many advanced ceramics are more stringent than some of those described in Part 2 of this book. One reason for the formation of the United States Advanced Ceramic Association was the need for manufacturers to work out new testing procedures suitable for advanced ceramics.)

To provide some insight into the spectrum of special characteristics and sophisticated applications of advanced ceramics, three important types — structural ceramics, electronic ceramics, and optical ceramics — will be discussed in some detail.

Structural ceramics

The term *structural ceramics* is applied to generally fine-grained, virtually pore-free materials that possess very high fracture strength, high fracture toughness, and great hardness. Such materials find increasing applications as intricately shaped, lightweight but strong components in engines and other machine components formerly made from metallic alloys. As such, they often must perform under dynamic loads, usually involving tensile stresses and impact loading, and frequently must function at high temperature in corrosive environments. These advanced structural ceramics should not be confused with traditional structural clay products which, while used as construction materials with important load-bearing implications, are relatively coarse-structured, fairly porous materials utilized for sustaining static compressive loads in relatively massive stationary structures in benign environments. Likewise, structural ceramics should not be confused with heavy refractories which are used primarily for static compressive load-bearing applications at elevated temperatures, although there is an increasing tendency to make spot use of structural ceramic materials in place of more conventional heavy refractories in certain stringent application zones in furnaces and process vessels.

The materials of major interest as structural ceramics are especially processed silicon carbide (SiC), silicon nitride (Si_3N_4), certain oxides (especially transformation-toughened zirconia [ZrO_2]), and a variety of ceramic-matrix composites.

In the case of silicon carbide and silicon nitride, the pure materials will not sinter under ordinary conditions, so they are frequently hot-pressed (or hot-isostatic-pressed), often with additives to provide a small amount of liquid to assist in bonding together the fine-grained matrix.

Several techniques have been developed for *near-net-shape forming* of these materials — that is, to fabricate powders into dense form without the extensive shrinkage that accompanies ordinary sintering and without requiring extensive machining following fabrication. So-called *reaction-bonded* silicon carbide is made by mixing silicon carbide powder with a relatively small amount of very fine carbon powder, pressing this mixed powder to shape, and then heating the compact while infiltrating it with molten silicon. The molten silicon reacts with the fine carbon in the interior of the compact to form additional silicon carbide, which serves to bond together the original silicon carbide grains. After cooling, unreacted silicon remains as a pore-filling second phase in the structure. Such a material is actually a composite of silicon carbide and silicon, and it makes a good structural ceramic for use below the melting temperature of silicon. Silicon nitride parts can also be near-net-shape fabricated by reaction bonding. Here, a mixture of silicon nitride powder and very fine silicon powder is pressed or otherwise shaped into a porous compact which is then heated (below the melting temperature of silicon) in a nitrogen atmosphere. The nitrogen penetrates into the pores, where it reacts with the fine silicon particles to form additional silicon nitride that acts to bond the powder into a strong mass. The final product has some residual porosity, but is quite suitable for many structural ceramic applications.

The toughening of zirconia — that is, making it more resistant to the propagation of small cracks under load, and therefore able to retain its structural integrity under loading conditions where an untoughened material would undergo catastrophic brittle failure—has resulted in this material being adopted for many advanced structural ceramic applications. The increase in toughness is imparted by uniformly distributed, very fine inclusions of an unstable (tetragonal) form of zirconia within the stabilized cubic zirconia matrix. When a microcrack encounters one of these inclusions, much of the energy of the propagating crack is absorbed in transforming the inclusion to its stable (monoclinic) form, and thus the crack is slowed or even stopped from further propagation. This mechanism of toughening with a dispersion of very fine nonstable zirconia inclusions can also be applied to other materials such as alumina ceramics. These strong, tough ceramics can be used for high-wear applications such as cutting tools, wire-drawing dies, and even (once sharpened) ceramic knife blades that remain sharp almost indefinitely. In addition to toughened oxides for wear-resistant applications, boron carbide (B_4C), ceramic-matrix composites, and even diamond are sometimes used.

A number of internal combustion engine parts are now being manufactured from structural ceramics, including rotors and stators for gas turbines or superchargers, piston head inserts, and valve heads and seats. In applications where monolithic ceramic parts have not yet found use, the metal parts may be coated with a ceramic "thermal barrier" to protect the underlying metal from heat and corrosion. In addition to their retention of strength at very high temperatures, structural ceramics are also much less dense than metals, so the great reduction in weight and inertia are considered as additional benefits of utilizing ceramic engine parts. There have even been suggestions that all-ceramic engines could operate without cooling (so-called "adiabatic" engines), which would greatly increase the thermal efficiency; much developmental work toward this goal is underway.

An especially demanding application of structural ceramics involves their use as armor, either for protecting personnel or vehicles. In this application, the key is ability to absorb the kinetic energy of a bullet or shrapnel particle by fracturing through the formation of a network of propagating microcracks. This not only protects the wearer or vehicle from penetration by the projectile, but the "gradual" nature of this cracking process also provides a considerable degree of shielding from the tremendous impact that would result from instantaneous conversion of the kinetic energy. Ceramic fiber-reinforced ceramic-matrix composite materials show great promise to replace monolithic structural ceramics for armor applications. Besides its shock-absorbing and projectile-stopping qualities, ceramic armor has the additional benefit of being lightweight compared to metal armor of comparable stopping power.

Certain ceramic materials can be used for structural applications in the human body, essentially as replacements for bone. Materials such as especially prepared alumina ceramics, vitreous carbon, synthetic apatite (calcium phosphate), and even certain glasses are biocompatible; the body will not only tolerate their implantation, but it will also incorporate them by the intergrowth of tissue into open pores. The necessity of gaining approval by the U.S. Food and Drug Administration and the long and complex clinical trials associated with the approval process have slowed the human implementation of bioceramic implants, but results of tests in animals are very promising.

Electronic ceramics

There are literally hundreds of applications of advanced ceramics that depend primarily on the reaction of the material to applied electric or magnetic fields. Some of these are enumerated here, along with a brief description of the special characteristics that make these materials useful for particular applications. In many cases, while the electronic properties are paramount, for many of these applications there are also stringent mechanical and thermal property requirements that must be met.

Many ceramic materials are electrical insulators and, consequently ceramics have been used for years for dc and low-frequency ac electrical insulator shapes ranging from large, high-voltage suspension insulators for power transmission lines to simple, low-voltage shapes for lamp and switch bases. These shapes have traditionally been made from clay-based porcelains and are not usually included in the advanced ceramics category. On the other hand, the utilization of advanced ceramic electrical insulator materials suitable for more exotic applications is growing very rapidly. The materials most often used are alumina ceramics, beryllia ceramics, aluminum nitride (AlN, a relatively recent addition), and a variety of special glasses, including those that can be converted into crystalline form after shaping (glass-ceramics). The most important electrical properties of such insulation materials are very low electrical conductivity, low dielectric constant (a low tendency to polarize or store charge), a high dielectric strength (resistance to breakdown under large voltage drops), and, for high-frequency applications, low dielectric losses (low propensity to convert energy in the alternating field into heat).

A wide variety of shapes are made from advanced ceramic insulator materials, many of which are so intricate that they must be made by injection molding or isostatic pressing followed by machining and finishing. An especially important ceramic insulation application is as smooth substrates for thick film and thin film deposition of circuitry. Substrates are usually made as thin (a few mm), flat rectangular sheets utilizing tape-casting technology. Most frequently, discrete electronic devices such as silicon chips or discrete capacitors will be attached to the film circuitry on the substrate to form what are known as hybrid circuits. It is not at all unusual for multilayer ceramic substrates to be employed. Multilayer substrates are made by thick-film printing of circuitry onto unfired ceramic tapes using metal inks, then stacking and laminating the green tapes together to form a sandwich structure, and then "cofiring" the ceramic and metal inks to form a single multilayer substrate. The circuits on different layers of the multilayer structure are connected at appropriate points by metal-filled holes called vias in the intervening ceramic layer. Substrates not only support the circuitry, but they also provide for dissipation of heat generated in the circuitry, either by absorbing it themselves or by conveying it to an attached heat sink. When substrates and their associated circuitry are fitted with external leads and are encapsulated to protect the circuitry from the environment, the entire assembly is usually called an electronic package.

Ceramic insulator materials are also commonly used as capacitor dielectrics — that is, as the material placed between the plates of a capacitor to serve as the charge storage medium. While any insulating material can be used for such an application, it is usually desirable to use materials that will allow the maximum amount of charge storage (capacitance) in the smallest possible device. This consideration means that materials with very high dielectric constants should be used. In addition to high dielectric constant, a capacitor dielectric should have high dielectric strength and low dielectric losses and should exhibit minimal variations in these properties with temperature or voltage changes. The most important group of advanced ceramic capacitor dielectric materials consists of combinations of barium titanate ($BaTiO_3$) with a variety of other oxides used to modify its fundamental properties. There are hundreds of titanate-based materials in use. Ceramic capacitor dielectrics are often made in the form of small, thin discs or thin-walled hollow tubes, with the plates being deposited on each side by thick film techniques.

A very important and rapidly growing form of high-rating ceramic capacitor, called a ceramic chip capacitor, is made by a process similar to that used for multilayer ceramic substrates. Very thin sheets of titanate dielectric are produced by tape casting, and a pattern of metal electrodes is thick-film printed onto one side. Many layers of tape are then stacked on top of one another and laminated together. Individual "chips" are diced out of this laminate and are fired to mature the ceramic-metal sandwich. These tiny chip capacitors can be soldered directly onto printed circuitry.

A number of ceramic materials are electrical insulators with respect to the movement of electrons, but nevertheless exhibit measurable electrical conductivities because of the ability of certain ions to move through the material when an electric field is applied. Such materials are called ionic conductors. If the conductivity is relatively high, they are called *fast ion conductors* or *solid electrolytes*. The most important fast ion conductors are AgI (Ag is the conducting ion), CaF_2 (F is the conducting ion), the so-called beta-aluminas (having roughly the formula $MAl_{11}O_{17}$, where M is silver or an alkali such as sodium, the M ion being the one responsible for conduction), zirconia (ZrO_2) doped with lime or yttrium oxide (with O being the conducting ion), and a number of special glasses (usually with alkali ions imparting conduction). Generally, the conductivity of ionic conductors increases rapidly with an increase in temperature, so they are almost always utilized at temperatures above room temperature, and sometimes at quite high temperatures. Their behavior as purely ionic conductors allows their use as solid electrolytes in high-temperature batteries and fuel cells, and the fact that only one particular type of ion moves in an electric field makes them useful as ion-specific sensor materials (an example is the use of stabilized zirconia as an oxygen sensor in automobile exhaust systems to sense the efficiency of the combustion process and activate changes in fuel-to-air ratios).

Although silicon, germanium, and gallium arsenide are the most utilized semiconductor materials, a number of other ceramic materials also are employed for semiconductor applications. Among the most used for these applications are various doped or slightly reduced oxides (especially ZnO) and doped silicon carbide. Such materials are commonly used as varistors (resistance changes with applied voltage) and thermistors (resistance changes with temperature). Varistors are commonly used to protect devices from damage by line surges (such as may be caused by lightning) by shunting current around the device when the varistor becomes highly conductive due to a voltage spike. Thermistors can be used as temperature measurement devices, and, if they are doped so as to have a positive temperature coefficient of resistance (their resistance increases with increasing temperature), they can be used as self-limiting heater elements in a variety of applications, including to rapidly heat automatic choke elements in automobile engines so that the choke quickly closes after start-up. When fabricated into single crystal form, ceramic semiconducting materials can be used to form pn junction diodes, and these can be used as power transistors, as light-emitting diodes (LEDs), and even as semiconductor laser diodes (the latter two devices are described in more detail in the upcoming section on optical or photonic ceramics).

A tremendous amount of interest has been focused on a group of special ceramic materials that function as superconductors at "high" temperatures. Superconductors have the unusual characteristics that they exhibit no electrical resistance and that they completely repel magnetic fields. (This latter effect is called the Meissner effect.) The first characteristic promises replacement of ordinary conductors to provide for much more efficient electrical power transmission and also the capability of building much more powerful magnets (which is the principal use of superconductors today). The Meissner effect has drawn great interest as a possible means of magnetic levitation of vehicles, especially trains, to virtually eliminate rolling friction and thereby allow for higher speeds and greatly reduced power consumption. All superconductor materials revert to "ordinary" behavior above a certain critical temperature. Between the discovery of the superconductivity phenomenon in 1911 and 1986, a number of metallic elements and compounds had been verified to show superconducting behavior, but all of them have extremely low critical temperatures, thus requiring cooling by liquified hydrogen (or even liquified helium) in order to remain superconducting; this requirement for bulky and expensive cooling systems has greatly limited the utilization of superconducting materials for practical applications.

Then, in 1986, a ceramic material (actually a complex oxide of copper, barium, and lanthanum) was discovered with a critical temperature as high as 40 K. The tremendous surge of interest in investigating similar ceramic compositions resulted in the discovery in 1987 of a material with a critical temperature as high as 90 K. Since that time, a number of superconducting ceramic oxides have been produced, most of which contain copper, an alkaline earth (such as Sr or Ba), and a rare earth (such as Y or La). Such "high-temperature" superconductors offer the possibility of cooling with inexpensive liquified nitrogen.

The pace of these discoveries has given hope that one day a superconductor material might be found with a critical temperature as high as room temperature (about 300 K), which would require no cooling at all to remain superconducting. Much research continues toward producing materials with higher and higher critical temperatures, but room temperature is still far above the highest critical temperature known at this writing. Most of the major potential uses of superconductor materials require that they be available in the form of fine wires, a form in which it is quite difficult to produce a brittle oxide material. Consequently, a great deal of the research on these new ceramic superconductors today is concentrated on novel fabrication technologies.

There are three families of advanced ceramics that find application because of their special magnetic properties; these materials are called *ferrites*. The *spinel ferrites* are known as "soft" magnetic materials, meaning that it is easy to reverse their magnetization direction with a small applied magnetic field. The spinel ferrites have the general formula $MO \cdot Fe_2O_3$, where M is a divalent transition metal ion such as Ni, Zn, Mg, Mn, Fe, or Co. Soft ferrites are usually used for very-low-loss transformer cores or inductors, where their low electrical conductivity inhibits heating by induced eddy currents, which commonly leads to losses with metallic materials. An especially important use of a spinel ferrite is as the particles dispersed in magnetic recording media such as magnetic tapes and computer memory disks. The *magnetoplumbite* or *hexagonal ferrites* are known as "hard" or permanent magnetic materials because they are magnetized by a very strong applied field during manufacture, and they will retain this strong magnetization permanently thereafter. Magnetoplumbite ferrites have the general formula $MO \cdot 6Fe_2O_3$, where M is primarily Ba, but with other divalent alkaline earth ions also substituting for some of the Ba. Hard magnetic materials are used for magnetic latches, for electric motors, for loudspeaker magnets, and for magnetic elements in ore separation processes. There is also some interest in hexagonal ferrite particles for high-density computer disk media. A third class of ceramic magnetic materials are known as the *garnet ferrites*. These materials have the general formula $3M_2O_3 \cdot 5Fe_2O_3$, where M is a rare earth such as Y or Gd; however, extensive substitution of other ions for M and for Fe lead to an ability to widely tailor the magnetic properties, a capability that has made these materials important in low-loss microwave applications. The most common composition in this series of materials is yttrium iron garnet, also called YIG.

A special group of ceramic insulator materials find application because they exhibit the *piezoelectric effect* — that is, when they are elastically strained, they generate a voltage, and when a voltage is applied, they undergo an elastic deformation. Such materials can be used for various electro-mechanical transducers, including sonar generators and detectors, ultrasonic cleaners, fixed-frequency resonators, phonograph and microphone pickups, and solid-state spark generators and ignitors. While many materials are known to show the piezoelectric effect in single crystal form (quartz being a common example), only the special group of ceramic materials called *ferroelectrics* can readily be made to show this type of behavior in polycrystalline (sintered) form. The most important ferroelectric ceramic materials are barium titanate ($BaTiO_3$), lead zirconate titanate (various alloys of $PbTiO_3$ and $PbZrO_3$, also designated PZT ceramics), and lanthanum-doped lead zirconate titantate (also designated PLZT ceramics).

Optical or photonic ceramics

The word *photonics* has been coined to describe the collective optical properties of materials and the application of these properties to make useful devices. This section begins with a brief description of some of the more important photonic properties of materials.

Wide band–gap materials, such as ceramic insulator materials, are inherently transparent to light in the range of wavelengths near to or including the visible range, provided that they do not contain internal inhomogeneities which can serve as scattering sites and which will reduce the transparency to translucency or even opacity. Consequently, ceramic single crystals, pore-free glasses and even pore-free single-phase polycrystalline ceramics can be utilized for photonic applications requiring transmission of light beams.

Even though a material is transparent, that does not mean that the material does not interact with light that passes through it. For example, the velocity with which light waves propagate varies from material to material, having its highest possible value in vacuum (this velocity being the universal constant $c = 3.00 \times 10^8$ m/second). In all other media, light travels slower than c, with the ratio of c to the actual velocity in the material being called the *index of refraction* of the material. On passing from one material into another having a different index of refraction, a light beam will bend; this principle is used when lenses cause light beams to focus or to diverge. In most materials, the index of refraction varies with the wavelength of the light; this behavior is called *dispersion,* and it underlies the separation of different wavelengths from a mixed light beam because of differing amounts of bending.

The electromagnetic waves that constitute a beam of light oscillate perpendicular to the direction of propagation of the beam. Under normal conditions, the oscillations are randomly oriented around the direction of propagation. However, some materials are able to modify the passing light beam so that only certain oscillation directions occur. This is called *polarization,* and many important applications of light require or take advantage of the polarization phenomenon. For example, if a light beam of a given polarization is incident on a material that will only transmit light of a different polarization, then that material will effectively block the passage of the beam.

Stress-free glass and many crystals are optically isotropic, meaning that the index of refraction is the same regardless of the direction of a light beam. However, certain types of crystals will split an incident light beam into two separate beams, each of which is polarized. Glass containing residual stresses will also show this type of behavior. This phenomenon is called *birefringence,* and materials exhibiting this behavior are said to be doubly refracting or optically active. This phenomenon is the basis for a number of types of optical devices and also as a means for revealing residual stresses in otherwise optically isotropic materials.

No material is perfectly transparent; some of the light entering a material will be absorbed and converted into heat or other forms of energy. Materials that do not absorb more strongly at one wavelength than at another appear to be colorless, but many transparent materials do show selective absorption and therefore appear to be colored.

Applied electric and magnetic fields can modify the refractive indices of materials to some degree; these effects are called the *electro-optic effect* and the *magneto-optic effect.* A particularly interesting ancillary effect of these phenomena is that an applied field can cause a normally optically isotropic material to display birefringence, which disappears when the field is removed. In certain materials, these effects are sufficiently large that optical devices can be built that take advantage of them.

The photonic applications of ceramic materials depend on one or a combination of the properties just described. For example, windows (used in the broadest sense to include parallel-sided thin slabs of material placed in openings into opaque chambers in optical instruments and devices as well as openings in buildings and vehicles) require simple transmission of light beams without alteration. On the other hand, filters are required to be transparent at certain wavelengths and to be strongly absorbing at others. A single material with the appropriate absorption characteristics can serve as a selective filtering window. Lenses are somewhat like windows except that they are made with surfaces that are not parallel to one another. Because a light beam is bent (refracted) as it passes into and out of the lens, it will exit in a different direction than it entered; thus, light beams can be manipulated to focus or defocus images.

An especially important photonic application of glass occurs in optical fibers. The function of these fibers is to carry a beam of light from one point to another without appreciable attenuation due either to absorption or to escape from the sides of the fiber. The most frequent application is in communications, where information is encoded in the form of modulations of the light beam, usually using a diode laser. Special glasses are most often used for optical fibers. To insure very low absorption, the glass must be extremely pure and free from inclusions. To prevent escape of the light beam from the sides, the fiber is usually made to have a central core of glass with low index of refraction, surrounded by a cladding of higher-index glass. The difference in index will cause perfect reflection of any portions of the beam that encounter the interface, thus insuring that all light launched within the core remains there no matter how the fiber may be curved.

Lasers are devices capable of producing highly energetic beams of light having all waves in phase and of the same wavelength. (The name *laser* is an acronym standing for *l*ight *a*mplification by *s*timulated *e*mission of *r*adiation.) Very-high-intensity lasers can be used for localized heating and melting, but certain types of lasers can also be used to produce very pure, modulated light signals and so are suitable for generating encoded beams used in optical fiber communications. To function, lasers must be "pumped"—that is, an input of energy is required to produce the unstable energy situation necessary for laser action to occur. Pulse lasers are usually pumped by means of an extremely bright flash of light; these lasers are often made from specially doped glasses or single crystals such as Cr-doped Al_2O_3 (ruby lasers) or Nd-doped yttrium aluminum garnet (YAG lasers). Other laser types can provide continuous output and so must be continuously pumped, often with electrical energy. An especially interesting type of continuous laser can be made from a semiconductor or insulator crystal that has been selectively doped so as to produce a pn junction. When a dc electrical voltage is applied across this junction in a direction that tends to force electrons toward the p side and holes toward the n side (forward biasing), recombination of excess electrons and holes in the junction region will release light energy, causing the junction to glow. When the electrical input is small, the light waves generated are not in phase, and the glowing junction is called a light-emitting diode (LED). LEDs are popular for constructing all sorts of electronic displays. When the electrical energy input is large, and certain other geometric requirements are met, the light emitted by the junction will be intense and in phase, the junction will behave as a laser. In lasing mode, the magnitude of the light emitted varies with the magnitude of the applied electrical signal, and the lasing behavior also "switches off" sharply when the pumping signal drops below a threshold level. Consequently, diode lasers are particularly well-suited for converting electrical signals into modulated light beams, and thus are especially valuable as signal generators in optical fiber communications systems.

At the receiving end of a fiber-optic communication system, there must be a detector capable of receiving a modulated beam of light and converting it back into an electrical signal. A pn junction diode will perform this function if a uniform dc voltage is applied in such a way as to tend to drive electrons toward the n side and holes toward the p side (reverse biasing). Light absorbed within the junction region of such a diode will generate free electrons and holes which will be swept away from the junction by the biasing voltage, resulting in an electrical signal proportional to the intensity of the light absorbed. Thus, the simplest fiber-optic communication system would have a diode laser with its output controlled by a modulated electrical pumping signal coupled to one end of the fiber and a diode detector coupled to the other end. In actual

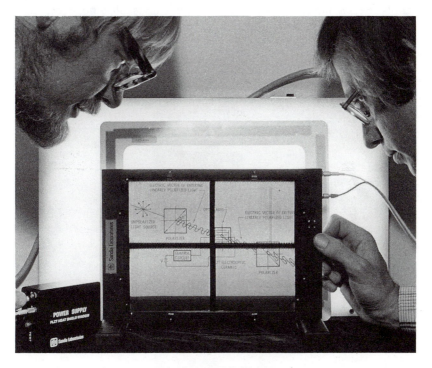

6.1. Employees in the Active Ceramic Materials Division of Sandia Labs demonstrate the high light transmission properties of an aircraft window segment developed for the U.S. Air Force to protect cockpit crews from burns and flashblindness. Segment can be switched from transparent (20% light transmission) to opaque (less than 0.01% transmission) in 50 microseconds (*courtesy Sandia Laboratories*).

practice, repeaters are also likely to be required at regular intervals along the fiber to correct for inevitable attenuation of intensity over long fiber lengths.

Transparent polycrystalline electro-optic materials, such as PLZT ceramics, can be used for a variety of devices in which transmission of a polarized light beam is throttled by changing the optical characteristics of the material with an applied electric field. The uses include rapidly darkening windows to shield pilots or other personnel from the intense flash of a nuclear explosion or a laser weapon (Figs. 6.1 and 6.2), goggles for welders, shutters for optical devices, optical displays, and even image storage devices. Whenever a polycrystalline ceramic is intended for use as a transparent material, very careful processing is necessary from starting powder through forming and firing in order to eliminate light-scattering pores and inclusions. It is not unusual to hot-press such ceramics in order to ensure the absence of porosity.

6.2. PLZT inserts in flight gear protect the wearer from flashblindness (*courtesy Sandia Laboratories*).

Testing 2 and calculation methods for industrial ceramics

PHYSICAL MEASUREMENTS AND CALCULATIONS are procedures used in the ceramic industry to maintain control of quality during production and to characterize the final product.

It is important to state at the outset that no attempt has been made to include the step-by-step procedures here for making these measurements. A great variety of standardized testing methods exists for nearly every common property, and it is important that one of these standards be selected by mutual agreement of the interested parties and then be rigidly adhered to. The importance of this rule is obvious when it is realized that using two different testing methods for the determination of the same property on the same specimen may very well give somewhat different results. Many firms use their own detailed testing methods, and others prefer to use methods adopted by the American Society for Testing and Materials (ASTM). Certain specialized trade organizations have adopted their own standards, some of which have eventually been adopted as ASTM standards. For example, the United States Advanced Ceramic Association is working within ASTM to establish standards for testing advanced ceramics.

The complete set of ASTM standards is reviewed, revised, and reissued each year (67 volumes in 1991 plus an index). ASTM standards are designated by code numbers such as C 182–88, where the letter and first number designate the specific standard, and the number following the dash is the last year of revision (in this case 1988). In some cases, a final year in parentheses indicates the last time the standard was reviewed (but not changed).

7

Elementary

BEFORE ENTERING into a discussion of elementary statistics, an explanation of accuracy, precision, and measurement units is in order. Whenever measurements and tests are performed, the two terms *accuracy* and *precision* invariably arise. The terms are not interchangeable. Accuracy is a measure of the closeness of a particular measurement to the true value of a property. If, for example, a metal rod has a true length at room temperature of 1.0000 m, then a measurement of the rod yielding 1.0002 m would be more accurate than one yielding 1.0004 m. The accuracy of a given measuring tool that has been properly calibrated is usually the smallest division or increment that the tool allows to be read. Even if it allows readings of small increments, however, a poorly calibrated tool will be inaccurate.

Precision refers to the reproducibility of multiple measurements of the same property. A poorly calibrated tool may give very reproducible measurement values which are far from the true value; the results from using such a tool would be precise but inaccurate. A tool yielding very reproducible values very near to the true value would be both precise and accurate.

Two systems of units are common in industry. The engineering system (English system) expresses lengths in feet or inches and weights in pounds or ounces. The metric system, or systeme internationale (SI), expresses lengths in meters and weights in grams. Except in the United States, laboratory workers throughout the world have traditionally used the metric system. Ceramic technologists must be prepared to work and think in terms of both systems. There is a conversion chart (Chart A.1)

statistics

in the Appendix to help convert from one system to the other when required.

Whenever a large number of measurements are made, as for example during quality control and quality assurance testing in a production plant, there is always some variation in the results of these measurements. This is true even when all items measured are from the same production run and all are tested in the same way. Part of this variation is a result of actual differences in the items being measured, and part is a result of errors made during the measurement.

A simple tabulation of the results of a large number of measurements of the same property is usually not sufficient to answer such important questions as: What is the single "best" value that describes this property? How much variation can be expected among individual items being produced and tested? Is the current production run producing items that are similar to or different from previous production runs? Is the overall production process operating under control, or has some important unexpected change occurred?

The usual route to answering these questions and others of similar nature is to first reduce the large volume of test data down to two or three representative quantities called statistics. The statistics for a set of data include the essential features and trends of the tabulated data, but are much easier to handle and compare than the data itself.

Two important references for persons involved in the testing of ceramic products are *ASTM Manual on Quality Control of Materials* (1951) and *ASTM Manual on Presentation of Data and Control Chart Analysis* (1976). The abbreviated treatment of statistics in this chapter is based, in part, on these manuals.

If the results of a large number of measurements of some property are graphed as frequency of occurrence of a specific value versus the

151

magnitude of the value, the result will usually resemble Figure 7.1. The bell-shaped curve is called a normal or Gaussian distribution and the properties of such a distribution curve are well known. Occasionally the curve may be slightly distorted or skewed, but usually this does not drastically affect statistics calculated for the distribution.

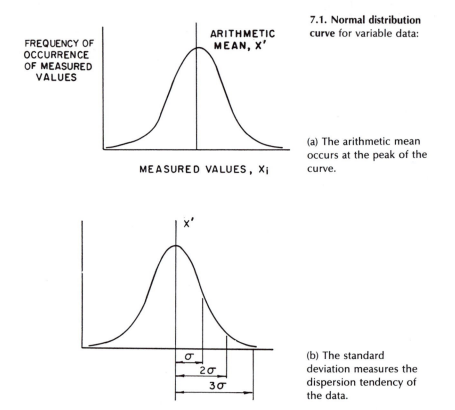

FREQUENCY OF
OCCURRENCE
OF MEASURED
VALUES

ARITHMETIC
MEAN, X'

MEASURED VALUES, X_i

7.1. Normal distribution curve for variable data:

(a) The arithmetic mean occurs at the peak of the curve.

(b) The standard deviation measures the dispersion tendency of the data.

In Figure 7.1, there is an obvious tendency for the measured values to cluster around some central "most frequent" value of the property. At the same time there is a tendency toward dispersion of the various measured values to produce the "sides" and outer "tails" on the distribution curve. It is possible to express both the clustering and dispersing tendencies of the data by calculating two simple statistics from the group of measured values.

The statistic used to express the most frequent value, or the centralizing trend, is the arithmetic mean, which is calculated according to

$$X' = \frac{\sum\limits_{i=1}^{n} x_i}{n} = \frac{x_1 + x_2 + x_3 + \ldots + x_n}{n} \tag{7.1}$$

where n is the total number of individual measurements. The value of X' represents the average of all the individual measurements x_i and normally corresponds to the value at the peak in the frequency distribution (Fig. 7.1[a]). The units on the mean are the same as those on the individual measurements.

The dispersion tendency of the data can be expressed by several different statistics, the most common of which is the standard deviation. This statistic is computed as

$$\sigma = \left[\frac{\sum\limits_{i=1}^{n} (x_i - X')^2}{n} \right]^{1/2}$$

$$= \left[\frac{(x_1 - X')^2 + (x_2 - X')^2 + (x_3 - X')^2 + \ldots + (x_n - X')^2}{n} \right]^{1/2} \tag{7.2}$$

The difference between each individual measurement (x_i) and the arithmetic mean X' is squared, these quantities are averaged, and the square root of this average yields the standard deviation. The units on the standard deviation are the same as the units on the individual measurements. (Note: Some treatments of statistics utilize $[n - 1]$ rather than n in the denominator of Equation 7.2. However, we follow here the practice of *ASTM Manual on Quality Control of Materials* [1951].)

The standard deviation is useful in describing the spread of the data because, provided that the data are normally distributed, this allows calculation of the percentage of all data points that should fall within specific upper and lower limits around the mean. For example (Fig. 7.1[b]), 68.3% of the data points should cluster between an upper limit of $(X' + \sigma)$ and a lower limit of $(X' - \sigma)$; 95.5% percent should cluster between the limits $(X' \pm 2\sigma)$; 99.7% should be found between the limits $(X' \pm 3\sigma)$. The significance of σ is thus clear: the smaller the value of σ for a set of data, the closer together individual values cluster in the data set.

The standard deviation is a measure of the precision (variability) of a set of data; the mean is a measure of the accuracy (approach to the true value) of the measurement technique. The American Society for Testing and Materials suggests that the minimum statistics to be reported for a set of data are the arithmetic mean (X'), the standard deviation (σ), and the number of individual data points (n). Many hand-held calculators and personal computer software programs provide for calculating these statistics for a data set.

In the statistical sense, a population is the whole class about which conclusions are to be made. All of the articles in a production run could constitute a population. Generally, quality control tests are not made on every item in a population, but rather are made on a small sample drawn from the population. If the sample is selected properly and is of sufficient size, the sample mean and sample standard deviation can be taken as good estimates of the population mean and standard deviation. The correct sample sizes for populations can be computed from probabilities based on the amount of risk of an improper quality decision that can be tolerated. A number of standard methods are available for sampling a population.

In complex multiple-step processes, such as those necessary for production of ceramic ware, some day-to-day or shift-to-shift variation in quality of the final product is inevitable. As long as the variation is within limits shown by experience to be typical, the overall process is said to be under control. As long as the variation is such that only a good product is produced under controlled conditions, the processes producing the product do not have to be changed.

Control charts are used to determine if a process is in or out of control. Control charts are chronological plots of some sensitive quality parameter (such as strength, density, or length) recorded each day, each shift, or at specific shorter intervals, including continuously. Figure 7.2 is an example of such a chart, plotting fired length of a nominal 9-inch refractory brick. The central line represents the expected mean value of the quality parameter over the long run. The upper control limit (UCL) and lower control limit (LCL) lines are established as the limits of probable variation in the parameter, based on experience. The example control chart indicates that the process was out of control on January 5, corrective measures were taken, and the process passed back into a controlled condition for the remainder of the time period. A control chart helps to make decisions about when corrective measures are needed, but only a thorough understanding of all steps in the production process can suggest where the problem is originating and what probable corrective measures are required.

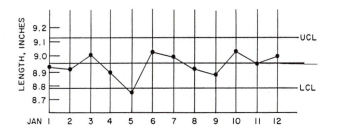

7.2. Control chart using length as the quality parameter. A length of 8.95 inches is the expected mean for this product. The process that produces this product was out of control on January 5.

Statistical process control, or SPC, is used to refine and re-engineer processes to bring them to and keep them within specified statistical control limits. The objective of SPC is prevention of occurrences of out-of-control situations rather than merely taking corrective action after the process has gone out of control. This approach can greatly reduce the cost of inspection, in-plant failures, and customer complaints. The additional cost of such prevention measures must be weighed against the deleterious effects of failures.

Some minor changes to improve a process can sometimes be made by factory personnel, but usually management must be foresighted enough to provide funding specifically for process improvement projects. Quality control personnel must watch for trends that may indicate that the process is likely to go out of control at some time in the future. Improvements should lead to better consistency by building quality into the process at the earliest possible stages of manufacturing.

Physical

CERAMIC PRODUCTS typically must meet dimensional, porosity, and density specifications designated by the customer or by industrial standards or governmental regulations. Since ceramic products undergo changes in all of these characteristics during processing, measurements of lengths, weights, and volumes must be made at various stages of the manufacturing process.

Length

At every step in the production of a ceramic product, the measurement of length comprises one of the most important quality control tests. All lengths are measured by comparing the dimension being measured to a standard length that has been carefully divided into a number of equal parts. The kind of tool used to measure length depends on the required accuracy of the measurement and to some degree on the shape of the object being measured. The size of the smallest division on a measuring tool determines its potential accuracy.

Because of normal variation, every article produced in a manufacturing process has slightly different dimensions. The amount of allowable variation in a dimension, the tolerance limit, is decided upon by agreement between the producer and the customer. It is generally unnecessary to measure length more accurately than the tolerance limit. Tolerance limits on lengths are often on the order of ½ to 1% of the nominal value, so that large objects can usually be measured with less accurate (and less expensive) tools than small objects.

The simplest and least accurate tool for measuring lengths is the

measurements

ordinary rule. Although they are often flexible, the most accurate rules are rigid. The smallest division normally found is 1/64 inch, but because of alignment errors when reading these rules, they probably are not reliable for measuring tolerances of less than 1/32 inch. Metric rules are available with the smallest division being 1 mm. Ordinary rules are of questionable reliability when used for measuring hole diameters and are usually completely unsatisfactory for measuring the depth of blind holes and depressions below a surface. The depth gauge, consisting of a rule and a flat-sided slide, is a tool expressly designed to make such measurements. The micrometer depth gauge shown in Figure 8.1 is a highly accurate instrument.

It is often inconvenient or even impossible to lay a rule along certain dimensions of an object. When this is the case, a measuring aid called a simple caliper can frequently be used. An inside caliper can be inserted into a hole and expanded until it contacts the sides. An outside caliper can be clamped around a part of the object. Both types of calipers can

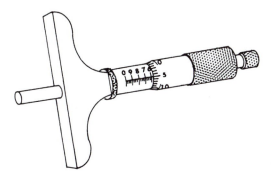

8.1. Micrometer depth gauge (*courtesy Hobar Publications*).

157

then be slipped free of the object and laid along a rule for measurement (Fig. 8.2).

8.2. Simple calipers (*courtesy Hobar Publications*):

A

(a) Simple calipers for outside (left) and inside measurements.

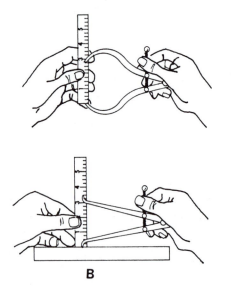

B

(b) Using a rule to measure the setting of outside and inside calipers.

A caliper may be incorporated as part of a rule, with one side fixed to the end of the rule and the other side free to slide along the rule. This kind of tool is called a slide caliper rule and is illustrated in Figure 8.3. The accuracy of such a tool depends upon the size of the smallest division on the rule, but it is inherently more reliable than a simple rule alone for measuring both outside and inside dimensions.

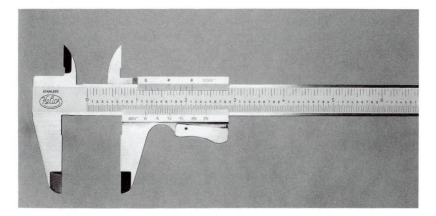

8.3. Slide caliper rule with verniers, which allow inside measurements (top caliper jaws) and outside measurements (bottom caliper jaws) (*courtesy Hobar Publications*).

The accuracy of a slide caliper rule can be increased by adding a vernier scale, which permits accurate division of the smallest inscribed division on the rule into even smaller divisions. The method for reading a vernier scale can be illustrated by an example. Figure 8.4 shows a portion of a slide caliper rule with two vernier scales (above and below the scales on the rule). The reading of the rule is determined by the location of the zero mark on the vernier scale. Each unit on the bottom scale of the rule corresponds to 0.025 inch. The bottom vernier divides each scale unit on the rule into 25 parts, or 0.001 inch. The reading on the bottom rule scale (zero mark on the vernier scale) is between 1.350 and 1.375 inches. The final reading is determined by locating the mark on the bottom vernier that aligns most nearly with a regular mark on the rule. The correct vernier mark in this case is 18, and therefore the reading on the rule scale is 1.350 plus 0.018, or 1.368 inches. Occasionally when the rule scale is divided into fractions of an inch (top rule scale), the vernier carries eight marks (top vernier) and can be read to the nearest one-eighth of the smallest rule division, and in this case, to the nearest $\frac{1}{128}$ inch. In Figure 8.4, the reading using the top rule and the top vernier is 1 $\frac{47}{128}$ inches.

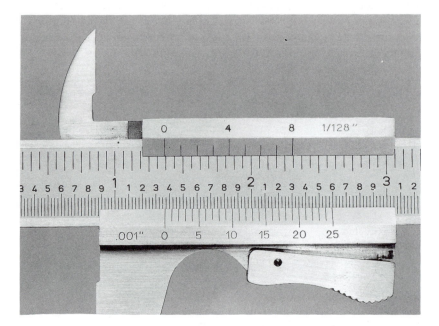

8.4. Close-up of verniers on a slide caliper rule. The readings on the bottom vernier is 1.368 inches and on the top vernier is 1 $\frac{47}{128}$ inches (*courtesy Hobar Publications*).

The most accurate routine measurements of length are usually accomplished with a tool called a micrometer caliper (or simply, a micrometer). Figure 8.5 shows an example of a micrometer. One side or jaw of the caliper (the anvil) is fixed, and the other (the spindle) is moved through the sleeve by turning a cylinder carrying very accurately machined threads. As the jaws of the caliper are screwed together, the beveled edge of the turning thimble moves along a scale inscribed on the sleeve. Each complete turn of the thimble moves it (and the spindle) a distance equal to one sleeve scale division. A second scale inscribed around the beveled edge of the thimble is divided into a number of equal parts. In this way, the smallest sleeve scale division is divided into a number of parts equal to the number of thimble divisions in one revolution. As an example, if the smallest sleeve scale division in Figure 8.5 is 0.025 inch and there are 25 divisions on the thimble scale, then dimensions can be measured to the nearest 0.025/25 = 0.001 inch. The read-

measuring surfaces · lock nut · ratchet stop

anvil · spindle · sleeve · thimble

knurled surface

spindle screw

—frame—

8.5. Micrometer caliper
(*courtesy Hobar Publications*):

(a) Parts of the caliper.

(b) Correct way of holding a micrometer caliper.

0.337

(c) Reading on the micrometer caliper is 0.337 inches.

ing of the micrometer is the sum of the reading on the exposed sleeve plus the reading on the spindle determined by the thimble mark that aligns with the scale on the sleeve.

Because of the possibility of damaging the fine threads on the spindle and thereby ruining the micrometer, these instruments should be handled with extreme care. The practice of tightening the spindle down hard onto the anvil is especially damaging to these devices.

It is often necessary to check the thickness of small parts as they move through a manufacturing process. The number and frequency of these measurements will likely require a measuring method faster than the use of a micrometer. An ideal tool for this rapid thickness testing is the dial gauge pictured in Figure 8.6. The movable plunger passing through the side of the gauge is linked to the dial indicator so that the gauge reads the length of the plunger movement directly. The gauge can be mounted in a jig with the plunger down in such a way that when the plunger rests on the jig base, the dial reads zero. Small parts can then be rapidly inserted between the plunger and the jig base, causing the plunger to slide into the gauge. The gauge reads the thickness of the part directly.

8.6. Dial gauge for measuring thickness of a glaze sprayed onto an unfired ceramic part (*courtesy Federal Products Corp.*).

Weight

The measurement of the weight of a ceramic article is often made as part of the determination of some other property such as density or porosity. Articles are also routinely weighed immediately after forming as a quality check in determining whether the forming process is running consistently. A most important type of weight measurement in ceramic technology occurs in the proportioning or batching of raw materials. The discussion of weighing in the present section is limited to laboratory measurements.

All weight-measuring instruments have a stated upper weight limit or capacity and are subject to serious damage if this limit is exceeded. This is particularly true of the delicate instruments used to measure small weights with high accuracy. Mechanical shocks and the presence of dust or corrosive vapors are just as damaging to many of the laboratory weighing instruments as exceeding their capacity would be. At the very least, mistreatment will cause these instruments to go out of calibration and may well necessitate expensive repairs or replacement. For these reasons, laboratory weighing instruments are kept covered when not in use, are usually mounted on level shockproof benches away from drafts and corrosive chemicals, and are always treated as delicate, expensive devices — which, in fact, they are. Even with careful handling, weighing devices may gradually go out of calibration, and they should be checked periodically using known weight standards.

Two principles are used in weight measurement, the spring principle and the balance principle. A weight attached to a vertical spring will stretch the spring a distance proportional to the weight. This is the principle on which spring scales are based. These devices are fairly insensitive to dust and vibration and for this reason can often be used in manufacturing areas. They are not high-accuracy instruments, however, and absolute accuracy generally drops off as the capacity increases.

In the balance principle of weighing, the weight to be determined is balanced with a set of known weights, using a pivoted beam. These devices, known collectively as balances, come in many sizes and modifications, running in capacity from large multiple-beam models capable of weighing a loaded railroad car down to those capable of measuring a few milligrams with an accuracy of 1 μg. Two balances are pictured in Figures 8.7 and 8.8.

The balancing lever pivots used in balances, called knife edges, carry the entire load of the object being weighed plus the standard weights plus the beam, and must remain free of friction and wear. For these reasons, especially in small, high-accuracy models, the knife edges are often made from polished hard materials such as agate or sapphire. The brittle nature of these materials requires that objects be placed gently on the pan to avoid damage to the knife edges. When an object is being weighed to the nearest milligram, both object and standard weights must be handled with tweezers to avoid errors due to fingerprints. To facilitate transfer of powders to and from laboratory balance

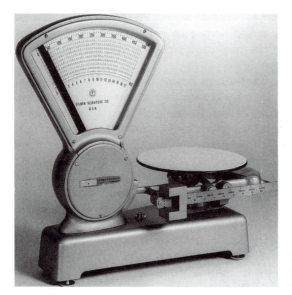

8.7. Balance with capacity of 10 pounds (*courtesy Fisher Scientific Co.*).

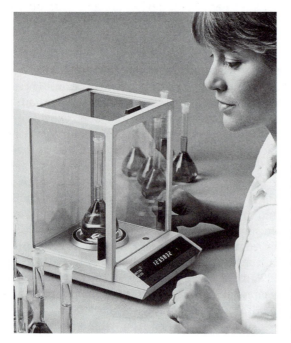

8.8. Sensitive analytical pan balance with a digital readout indicator. The delicate pivoted beam arrangement is located in the enclosure below and behind the pan compartment. This instrument has a maximum capacity of 205 g and reads to the nearest 0.1 mg (*courtesy Mettler Instrument Corp.*).

pans as well as to prevent corrosion of the balance pans by reactive materials, glassine weighing paper is usually used to hold the sample. Very sensitive balances give incorrect readings if air currents are moving around the pans, so they are usually enclosed in a glass case which should be kept closed during weighing (Fig. 8.8). Hot samples should never be placed on the pans of such balances, since the thermal expansion of the balance parts plus the convection currents arising in the enclosure will cause an incorrect weighing.

Microprocessor-controlled weighing equipment and electronic balances utilizing strain gauges or force cells instead of standard balancing weights are common in ceramic factories and laboratories. These instruments must still be calibrated with standard weights. The accuracy of the device will not exceed the accuracy of the calibration weights. Taring — that is, compensation for the weight of sample containers or weighing papers — can be done automatically with most electronic balances, and this is a distinct advantage over conventional balances.

Volume

The measurement of the volume of ceramic articles is a routine laboratory procedure as part of the determination of density and porosity. At first glance it would seem that the obvious way to determine the volume of an object would be to measure its dimensions and then to calculate the volume from known geometric formulas. This procedure would be satisfactory if the object had a simple geometric shape such as a perfect rectangular solid or cylinder. Most ceramic articles, however, are too complex in shape to allow this simple approach, and a different volume measurement method must be used.

An indirect way of measuring the volume of a solid object utilizes fluid displacement. If a container is completely filled with some inert liquid (perhaps water), and a nonporous solid object is then placed carefully on the surface of the liquid and allowed to sink, some of the liquid will spill out over the lip of the container. The volume of the liquid displaced by the object will be exactly equal to the external volume of the immersed object. If some means was provided for catching all the displaced liquid, it could be weighed and the volume of the solid object could easily be calculated from the known density of the liquid.

$$\text{Solid volume} = \text{volume of liquid displaced} = \frac{\text{weight of liquid displaced}}{\text{density of liquid}} \qquad (8.1)$$

Similarly, the rise of liquid in a partially filled graduated volumetric cylinder can be directly observed as a solid is submerged in the liquid.

All objects exhibit buoyancy—that is, an apparent reduction in weight when placed in a fluid—and this effect is commonly used to measure the volume of ceramic objects. Archimedes' principle relates the apparent reduction in weight of a submerged object to the volume of that object and the density of the liquid. This principle states that "an object placed in a fluid suffers a loss in weight equal to the weight of the fluid which the object displaces." Thus, if an object is completely submerged in a fluid, it will lose an amount of weight equal to its volume multiplied by the density of the fluid. (This principle applies whether the submerging fluid is a liquid or a gas, but the effect is much greater for a liquid.)

Loss in weight of solid when submerged

$$= \text{volume of fluid displaced} \times \text{fluid density}$$
$$= \text{volume of solid} \times \text{fluid density} \qquad (8.2)$$

so that

$$\text{Volume of solid} = \frac{\text{loss in weight of solid when submerged}}{\text{fluid density}} \qquad (8.3)$$

This formula (which is later modified to Equation 8.10 for porous ceramic solids) is the basis for most of the volume measurements made on complex ceramic shapes. Special balance arrangements involving a hanging submersible sample pan are used for measuring the weight loss.

When using Archimedes' principle, several things must be kept in mind. First, the density of any liquid changes slightly with temperature, and for very accurate work, the liquid density corresponding to the actual laboratory temperature must be used in calculations. The correct density value for many common liquids can be found in handbooks such as Perry and Chilton (1973). The second thing to consider when using loss in weight to measure volume is that the submerged portion of the balance pan and its suspending support will also lose some weight. This problem can be overcome by setting the balance to zero with the balance pan submerged to the same depth as when it is actually making a submerged weighing. Finally, any bubbles trapped below the balance pan or between the pan and the object during a submerged weighing will introduce considerable error (erroneously large values of volume), and for this reason, the balance pan is often replaced by a wire sling. A small amount of wetting agent such as a detergent added to the submerging liquid will help to prevent problems with entrapped bubbles.

Six different volumes can actually be ascribed to a ceramic object. The reason for this complexity is the presence of porosity in most ceramics. Figure 8.9 shows schematically the kinds of volumes in a porous object. Some pores will likely be connected to the surface, such as B in Figure 8.9. These pores are called *open pores,* and the total volume of these open pores in an object is called the *open pore volume,* usually written V_{op}. There will very likely also be some pores such as C in the object which are not connected to the surface. These are called *closed pores,* and the total volume of these pores is the *closed pore volume,* V_{cp}. The sum of the closed pore volume and the open pore volume is the *total pore volume,* V_p.

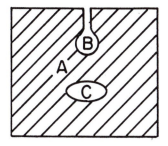

8.9. Schematic of types of volumes in a ceramic body: A = solid; B = open pore; C = closed pore.

The volume actually occupied by solid material in the object (represented by the shaded material A in Fig. 8.9) is called the *true volume* of the object and is written V_t. The volume of the solid material plus the closed pores ($A + C$) is the *apparent volume* of the object, V_a. Finally, the volume that would be computed if the external dimensions of the object were measured and appropriate geometric formulas were applied is the *bulk volume* of the object, V_b. The bulk volume includes not only the actual solid material, but also all of the pores, both open and closed.

The following relationships exist between the six volumes described above:

$$V_p = V_{op} + V_{cp} \tag{8.4}$$

$$V_b = V_t + V_p = V_t + V_{op} + V_{cp} \tag{8.5}$$

$$V_a = V_t + V_{cp} = V_b - V_{op} \tag{8.6}$$

$$V_t = V_b - V_p = V_a - V_{cp} \tag{8.7}$$

$$V_{op} = V_b - V_a \tag{8.8}$$

$$V_{cp} = V_b - V_t - V_{op} = V_a - V_t \tag{8.9}$$

The values of V_b and V_{op} can be measured directly and nondestructively on the object. The value of V_t can be measured only if the object is first crushed into a fine powder. (See measurement of true density, next section.) The value of V_{cp} can never be measured directly, but must be calculated by one of the relationships shown above.

The bulk volume is the one most often measured, and when the volume of an object is mentioned with no qualifying adjectives, it is almost always the bulk volume that is referred to. Several of the volumes discussed must be determined when sample porosity and density are to be calculated (see following sections). In the case of porous ceramics, it is necessary to first completely fill the open pores with liquid before a submerged weight is taken. This pore-filling procedure, called saturation, is usually accomplished by boiling the object in the liquid or by evacuation of air from the open pores followed by immersion and a long soaking period.

Using the symbols W_D for the unsaturated (dry) weight of the object, W_S for saturated weight (weight "in the open" but with all open pores filled with liquid), W_{ss} for the weight of the saturated object when it is submerged in the liquid, and ϱ_L for the density of the saturating and submerging liquid, the following relationships hold:

$$V_b = \frac{W_S - W_{ss}}{\varrho_L} \tag{8.10}$$

$$V_{op} = \frac{W_S - W_D}{\varrho_L} \tag{8.11}$$

$$V_a = V_b - V_{op} = \frac{W_S - W_{ss}}{\varrho_L} - \frac{W_S - W_D}{\varrho_L} = \frac{W_D - W_{ss}}{\varrho_L} \tag{8.12}$$

Density

The density of any material, be it solid, liquid, or gas, is defined as the weight per unit volume. The units on this quantity are usually either lb/ft³ or g/cm³. Density is probably the most often measured and reported property for all types of ceramics. Since the calculation of density requires that the weight of a sample be divided by its volume, the measurement of volume constitutes an important part of any density determination.

The application of the relationships in Equations 8.10–8.12 allows most of the routinely used densities and porosities to be determined with three simple weight measurements (W_D, W_S, and W_{SS}).

The density most commonly determined for a ceramic object is the *bulk density*, ϱ_b, which is the weight divided by the bulk volume. Whenever a density is reported that carries no further specification, it is usually safe to assume that bulk density is meant. The bulk density can be calculated by applying Equation 8.10 to give

$$\varrho_b = \frac{W_D}{V_b} = \frac{W_D \times \varrho_L}{W_S - W_{SS}} \tag{8.13}$$

In a similar manner, the *apparent density*, ϱ_a, is based on the apparent volume and is calculated as

$$\varrho_a = \frac{W_D}{V_a} = \frac{W_D \times \varrho_L}{W_D - W_{SS}} \tag{8.14}$$

Finally, the *true density* of an object, ϱ_t, is based on the true volume and is calculated as

$$\varrho_t = \frac{W_D}{V_t} \tag{8.15}$$

The determination of the true volume and true density of an object requires a great deal more effort and time than the bulk and apparent properties. The traditional method requires a finely ground sample (in order to eliminate the presence of closed porosity) and a special type of volumetric bottle called a pycnometer. The procedure requires several careful weighings and is subject to considerable error if not carried out properly. The details of the procedure are given in ASTM standard C 135–86.

8.10. Gas pycnometer for rapid determination of true volumes (*courtesy Quantachrome Inc.*):

(a) View of the instrument.

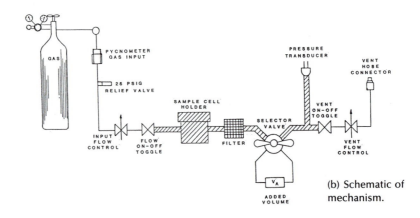

(b) Schematic of mechanism.

A more modern method for determining true volume and true density of a powder uses a gas pycnometer such as the one shown in Figure 8.10 (a). Two precision cells of known volume are an essential part of the apparatus (Fig. 8.10 [b]). There is a pressure release valve between the cells. In typical use, a weighed sample of powder is placed in the first cell, with the second cell at ambient pressure. The pressure in the first cell is increased to about 15 psi over ambient, then the relief valve is opened, equalizing the pressure between the cells. The ideal gas law is assumed to hold, and this allows the powder true volume to be calculated as

$$V_t = \frac{V_C + V_A}{\left(1 - \dfrac{P_2}{P_3}\right)} \qquad (8.16)$$

where

V_t = true volume of the powder under test
V_C = volume of the powder test cell when empty
V_A = volume of pressure release cell
P_2 = the increased test cell pressure, usually 15 psi
P_3 = the final pressure in the system after the relief valve has been opened

With such an instrument, the true volume and therefore the true density of powders can be determined accurately to within 0.1 to 0.2%. Helium is generally used as the pressuring gas to approximate ideal gas behavior. However, other gases such as nitrogen may be used in most situations without introducing major error.

Specific gravity

The specific gravity (SG) of a material is simply the ratio of the density of the material to the density of pure water. Since both quantities in the ratio carry the same units, the specific gravity is a unitless quantity. Except where great accuracy is necessary, it is usually satisfactory to take the density in g/cm^3 as being numerically equal to the specific gravity.

The step-by-step procedure used in measuring densities, especially the saturation step, can influence the final result somewhat. Therefore it is always necessary to carry out these measurements according to some standardized procedure so that all results will be comparable. Individual company standard test methods are sometimes used, but the most common practice is to follow one of the several available ASTM Standard Methods of Test. A partial list of these standards used for various types of ceramic products is given in Table 8.1.

TABLE 8.1. ASTM standard test methods for density of ceramics

ASTM Designation	Applicable Material
C 20–87	Burned refractory brick
C 493–86	Granular refractory material
C 357–85	Granular refractory material
C 134–84	Rectangular refractory brick
C 135–86	True density of refractory materials
C 329–88	Fired whitewares
C 373–88	Fired whitewares
C 559–90	Carbon and graphite articles

A fairly common practice in reporting the density of a material is to express it as a ratio of the bulk density to the true density. When reported in this way, it is called the *percent theoretical density* and is given by

$$\% \text{ theoretical density} = \frac{\varrho_b}{\varrho_t} \times 100 \tag{8.17}$$

Porosity

The porosity of a ceramic, particularly a fired ceramic, is usually a very carefully controlled property. The greater the porosity of a material, the more likely will be the penetration of liquids and vapors into the material during use, and such penetration usually is accompanied by potential structural damage to the material. Thus, porous building brick and tile are subject to water penetration and resulting damage from freeze-thaw cycling. Refractories with high porosity will suffer internal chemical attack as a result of the penetration of slags or molten metals into the interior. Tableware of high porosity, if not glazed, would absorb various substances during use and become permanently stained and unsanitary. In a few cases, especially when producing thermal insulation or filtering media, a high porosity is desirable, but the amount and size of pores must still be controlled to ensure adequate strength. There are few ceramic products produced today which do not have very carefully controlled (and therefore carefully measured) pore structures.

Only the open pore volume of a specimen, also called the apparent pore volume, can be directly measured. When this volume is expressed as a percentage of the bulk volume of the specimen, it is called either the *percent apparent porosity* or, most often, simply the *porosity*.

$$\% \, P_a = \frac{V_{op}}{V_b} \times 100 \tag{8.18}$$

(A possible point of confusion is that the calculation of percent apparent porosity is based on the object's *bulk* volume and not on its *apparent* volume.)

When the appropriate weight quantities are substituted from Equations 8.10 and 8.11 in Equation 8.18, the result is

$$\% \ P_a \ = \ \frac{W_s \ - \ W_D}{W_s \ - \ W_{ss}} \times 100 \tag{8.19}$$

Another useful relationship for calculating apparent porosity involves the bulk and apparent densities of the object:

$$\% \ P_a \ = \ \frac{\varrho_a \ - \ \varrho_b}{\varrho_a} \tag{8.20}$$

A property very closely related to the apparent porosity of a specimen is its *percent water absorption*. This is the ratio of the weight of water absorbed during saturation to the weight of the specimen when it is unsaturated. If the symbol W_{ws} indicates the weight of the water-saturated specimen, the percent water absorption is given by

$$\% \ A \ = \ \frac{W_{ws} \ - \ W_D}{W_D} \times 100 \tag{8.21}$$

The closed pore volume, and thus also the total pore volume, must be determined indirectly. These quantities would be used to compute *percent closed porosity:*

$$\% \ P_c \ = \ \frac{V_{cp}}{V_b} \times 100 \tag{8.22}$$

and the *total porosity*

$$\% \ P_t \ = \ \frac{V_{op} \ + \ V_{cp}}{V_b} \times 100 \tag{8.23}$$

Also

$$\% \ P_t \ = \ \frac{\varrho_t \ - \ \varrho_b}{\varrho_t} \times 100 \tag{8.24}$$

and

$$\% \ P_t \ = \ \left(1 \ - \ \frac{\% \ \text{theoretical density}}{100}\right) \times \ 100 \qquad (8.25)$$

The procedure used for measuring porosities and water absorption should be one selected from the many available as ASTM standards or other well-documented sources (Table 8.2). Whenever results are reported, a note should be included specifying the testing procedure used.

TABLE 8.2. **ASTM standard test methods for porosity and absorption of ceramics**

ASTM Designation	Applicable Material
C 20–87	Burned refractory brick
C 493–86	Granular refractory materials
C 373–88	Fired whitewares
D 116–86 (1990)	Vitrified electrical ceramics
C 301–90	Clay pipe
C 67–90a	Building brick

Several additional procedures for measuring porosity are available which do not utilize the conventional saturation and submersion scheme. The first, *intrusion porosimetry,* uses the fact that a nonwetting liquid will not penetrate into a small pore unless an external pressure is applied. Mercury is the nonwetting liquid most commonly used, and higher and higher pressures are needed to force mercury into smaller and smaller pores. If the amount of mercury that has been forced into a sample has been recorded as a function of pressure applied, not only the open pore volume but also the size distribution of the open pores can be determined. A second method, *permeametry,* uses the flow rate of a gas through a porous material in calculating the volume of open, interconnected pores as well as their diameters. (Permeametry does not detect open pores that do not extend completely through the specimen.) Various gas adsorption methods require special apparatus for first adsorbing a certain gas on the interior surfaces of the open pores and then stripping it away. Such measurements are usually highly automated.

Shrinkage

Ceramic objects undergo shrinkage at several different points in the manufacturing sequence. While the general shape of a ceramic object changes little during processing (small amounts of warpage are usually unavoidable), the size of a finished piece is almost always significantly smaller than the size of the newly formed piece. During drying of a freshly formed piece, the individual grains in the ceramic article draw together until they touch, resulting in drying shrinkage. The greater the amount of water present in the formed piece, the greater will be the amount of drying shrinkage. Further shrinkage occurs during the firing operation as the pores are gradually reduced or eliminated. This is the firing shrinkage. The linear shrinkage — that is, the amount by which each dimension changes — is more commonly reported than the volume shrinkage — the amount by which the volume of the whole article changes. The two are directly related. Since the fired dimensions are those of ultimate importance for application, the amounts of both linear drying and linear firing shrinkage must be known and must be controlled so that forming dies or molds with the proper amount of oversize may be prepared.

Some confusion can arise in reporting shrinkages unless meticulous care is taken to also report the basis on which the calculation has been made. Thus, if L_f is an as-formed dimension and L_d is the same dimension of the dried unfired piece, *percent linear drying shrinkage on the formed basis* is given by

$$\% \ LDS_f \ = \ \frac{L_f \ - \ L_d}{L_f} \ \times \ 100 \tag{8.26}$$

On the other hand, *percent linear drying shrinkage on the dried basis* is given by

$$\% \ LDS_d \ = \ \frac{L_f \ - \ L_d}{L_d} \ \times \ 100 \tag{8.27}$$

The *percent volume drying shrinkages* are given by

$$\% \ VDS_f \ = \ \frac{V_f \ - \ V_d}{V_f} \ \times \ 100 \tag{8.28}$$

and

$$\% \ VDS_d \ = \ \frac{V_f \ - \ V_d}{V_d} \ \times \ 100 \tag{8.29}$$

where all volumes are bulk volumes.

There is usually no confusion in reporting firing shrinkages, since traditionally they are based on dried unfired dimensions.* If L_F represents the fired dimension, *percent linear firing shrinkage* is given by

$$\% \ LFS \ = \ \frac{L_d \ - \ L_F}{L_d} \times 100 \tag{8.30}$$

The *percent volume firing shrinkage* is given by

$$\% \ VFS \ = \ \frac{V_d \ - \ V_F}{V_d} \times 100 \tag{8.31}$$

Note that the following very useful relationships exist between formed and fired dimensions:

$$L_f \ = \ \left(\frac{L_F}{1 \ - \ \% \ LFS/100} \right) \left(\frac{\% \ LDS_d}{100} + 1 \right)$$

$$= \ \frac{L_F}{(1 \ - \ \% \ LFS/100) \ (1 \ - \ \% \ LDS_f/100)} \tag{8.32}$$

The simplest way to determine linear shrinkages is to make a measurement of the dimensions before and after the shrinkage occurs. Sometimes it is not convenient to make length measurements, but displacement volumes for the object can readily be determined. Very simple geometric relationships exist between volume shrinkages and linear shrinkages. For drying shrinkage,

$$\% \ LDS_d \ - \ \left[\left(\frac{\% \ VDS_d}{100} + 1 \right)^{1/3} - 1 \right] \times 100 \tag{8.33}$$

$$\% \ LDS_f \ = \ \left[1 \ - \ \left(1 \ - \ \frac{\% \ VDS_f}{100} \right)^{1/3} \right] \times 100 \tag{8.34}$$

For firing shrinkage,

---

* There is a trend, particularly among technical ceramic manufacturers, to express firing shrinkage as $\% LFS_F = [(L_f - L_F)/L_F] \times 100$, incorporating both drying shrinkage and firing shrinkage into one expression. This can be simplified to $\% LFS_F = [(L_f/L_F) - 1] \times 100$ where the ratio L_f/L_F is called the shrinkage factor. This factor is of particular convenience in designing forming dies of the proper dimensions to produce a desired fired size. It can be seen that any fired dimension, when multiplied by the shrinkage factor, will yield the required freshly formed dimension.

$$\% \ LFS \ = \ \left[1 \ - \ \left(1 \ - \ \frac{\% \ VFS}{100} \right)^{1/3} \right] \times \ 100 \qquad (8.35)$$

These relationships assume that linear shrinkage is uniform in all directions, a condition sometimes not met because of preferred orientation of irregularly shaped particles that occurs during forming.

9

Batch

CERAMICS ARE COMPOSED of complex, manufactured minerals of controlled physical and chemical composition. Most ceramics begin as mixtures of various raw materials which, when heated, decompose and react with one another to yield the final fired product. This chapter is concerned with the calculations necessary to determine what blend of raw materials should be used to achieve a required fired chemical composition. These calculations are absolutely essential in the production of glasses and the compounding of glazes and enamels. They are also of considerable use in compounding ceramic bodies, but here the physical and mineralogical properties of the raw materials must also be considered in addition to their chemical compositions when determining the appropriate blend. For example, a ball clay and a china clay may have essentially the same chemical composition, but their physical properties are very different, and substituting one for the other would greatly influence the tempering, forming, and drying behavior of a body.

Oxide formulas and formula weights

The greatest proportion of all ceramics can be thought of as complex oxides, and the raw materials used to produce them are also generally complex oxides. It is convenient to write the chemical formulas of such materials as *oxide formulas* of the general form

calculations

$$\begin{matrix} RO \\ R_2O \end{matrix} \cdot R_2O_3 \cdot RO_2 \cdot H_2O \qquad\qquad (9.1)$$

where R represents metallic elements. The RO group consists of alkaline earth oxides such as CaO, MgO, and BaO; the R_2O group includes the alkali oxides such as K_2O, Na_2O, and Li_2O; the R_2O_3 group contains the sesquioxides such as Al_2O_3 and Fe_2O_3; and the RO_2 group consists of the dioxides, the most important of which is SiO_2. The addition of H_2O on the end of the oxide formula allows hydrated minerals and hydroxides to be expressed in this form. Sometimes the RO and R_2O oxides are grouped together and called *basic* oxides, the R_2O_3 oxides are called *neutral* oxides, and the RO_2 oxides are called *acidic* oxides.

As an example of the expression of the composition of a mineral as an oxide formula, consider the pure clay mineral kaolinite, usually given the theoretical chemical formula $Al_2(Si_2O_5)(OH)_4$. When rewriting this as an oxide formula, it is first noted that, since no alkali or alkaline earth elements are theoretically present, there will be no RO or R_2O parts to the oxide formula. The two aluminums combine with three of the nine oxygens to form one Al_2O_3 unit. The two silicons combine with four of the six remaining oxygens to form two SiO_2 units. The remaining two oxygens combine with the four hydrogens to form two H_2O units. The oxide formula for kaolinite is thus $Al_2O_3 \cdot 2SiO_2 \cdot 2H_2O$; the absence of a number in front of the Al_2O_3 is taken to mean that there is only one of these units. Several other examples of the expression of chemical formulas as oxide formulas are given in Table 9.1.

It is important to realize that these materials do not really consist of mixtures of the various simple oxides, but rather that the oxide formula

TABLE 9.1. **Examples of oxide formulas**

Material	Chemical Formula	Oxide Formula
Calcite or limestone	$CaCO_3$	$CaO \cdot CO_2$
Sodium bicarbonate	$NaHCO_3$	$Na_2O \cdot 2CO_2 \cdot H_2O^a$
Borax	$Na_2B_4O_7 \cdot 10H_2O$	$Na_2O \cdot 2B_2O_3 \cdot 10H_2O$
Ilmenite	$FeTiO_3$	$FeO \cdot TiO_2$

[a]The chemical formula must be multiplied by two before the oxide formula can be written in this case.

is simply another way of expressing the total chemical composition of the mineral. Later in this section, a method is given for calculating the oxide formula of a mineral from its chemical analysis.

It is necessary that a *formula weight* be assigned to each material. This consists of the sum of the molecular weights of each simple oxide in the oxide formula multiplied by the number of such units in the formula. The molecular weight of an oxide unit is equal to the sum of the weights of all the atoms in the simple oxide unit:

Formula weight of calcite ($CaCO_3$)
  = (mol wt CaO) + (mol wt CO_2)
  = 56.1 lb + 44.0 lb
  = 100.1 lb*

Formula weight of sodium bicarbonate ($NaHCO_3$)
  = (mol wt Na_2O) + 2(mol wt CO_2) + (mol wt of H_2O)
  = 62.0 lb + 2(44.0) lb + 18.0 lb
  = 168.0 lb

Notice that the formula weight of a compound is not necessarily the same as the molecular weight, although it is always a multiple of it. Thus the formula weight of $CaCO_3$ equals its molecular weight, and the formula weight of sodium bicarbonate is twice its molecular weight. Unless otherwise indicated, the formula weight will be used throughout this chapter in lieu of the molecular weight. Table A.1 (see Appendix) lists formula weights of common oxides and theoretical minerals.

---

*Molecular weight can be expressed in grams (gram molecular weight) or in pounds (pound molecular weight). Both have the same numerical value but, of course, represent very different quantities of material. In the batch calculations in this chapter, pound molecular weights and pound formula weights will be used, so the units on these numbers will be pounds.

Loss on ignition and moisture

When a raw material is heated in air, it often loses weight as a result of volatiles being driven off. In general, any material whose oxide formula contains H_2O, CO_2, SO_2, or NO_2 units will lose weight if heated to a sufficiently high temperature. Chemically combined H_2O, CO_2, SO_2, and NO_2 come off irreversibly at specific temperatures for each material. On the other hand, water of hydration (molecular water such as that in the mineral borax) is lost gradually with heating. Any carbonaceous material present combines with oxygen from the air to form volatile oxides which also leads to weight loss.

The amount of weight lost by a material during heating is an important characteristic of the material. This weight loss, called the *loss on ignition* (LOI), is readily determined by experiment or may be calculated for a material if its exact oxide formula is known. While specific and detailed procedures are available for determining LOI (see for example ASTM C 573–81[1990]), they all consist of essentially the same steps. A small representative sample is first dried to constant weight W_D at 110°C (230°F) and then ignited or heated until it reaches constant weight W_I at a specific temperature. The LOI is then computed as

$$\% \text{ LOI} = \frac{W_D - W_I}{W_D} \times 100 \tag{9.2}$$

The procedure of igniting or heating a ceramic material to drive out volatiles is also called *calcination*. A material loses some weight during the drying step, which precedes calcination. The amount of this drying weight loss depends on current weather conditions (especially humidity) and the previous method of storage of the material. The drying loss is not included in the %LOI, but it must be known to determine the weight of undried material necessary to give a certain weight of dry material called for in a recipe (or batch formulation). This moisture content can be expressed on either the undried or the dried basis, and it is necessary to specify the basis when moisture content is being reported. On the undried basis

$$\% \text{ moisture } (u) = \frac{W_u - W_D}{W_u} \times 100 \tag{9.3}$$

where W_u is the undried (stored or as-received) weight. On the dried basis

$$\% \text{ moisture } (D) = \frac{W_u - W_D}{W_D} \times 100 \tag{9.4}$$

When it is necessary to compute the amount of undried moist material required to yield a certain amount of dry material, the following expression will be useful:

$$W_u = W_D \left[1 + \frac{\% \text{ moisture } (D)}{100} \right] = \left[\frac{W_D}{1 - \% \text{ moisture } (u)/100} \right] \tag{9.5}$$

Empirical formula and formula weight

The procedure below is useful for computing the oxide formula from the chemical analysis of a glass or fired ceramic or of a raw material that does not contain volatiles. The same procedure is applicable to materials containing volatiles, provided that the chemical analysis is complete enough to include amount and type of volatiles. Since this formula is computed from an experimental or empirical analysis of the mineral, it is called an *empirical formula*.

1. All calculations are based on 100 pounds of mineral or ceramic so that the weight percentage of each oxide in the chemical analysis becomes the weight in pounds of that oxide in the 100-pound sample. The first step is to list the weight of each oxide.
2. The weight of each oxide is divided by its respective molecular weight to yield the number of moles of each oxide.
3. These mole numbers are normalized by one of the following methods:
   A. If the ceramic is a glass, glaze, or enamel, or if the mineral is a feldspar, the number of moles of each oxide (step 2) is divided by the sum of the number of moles of RO plus R_2O oxides.
   B. If the ceramic is a body or if the mineral is a clay, the number of moles of each oxide (step 2) is divided by the number of moles of alumina (Al_2O_3).
4. The normalized number of moles of oxides are arranged in the form of a standard oxide formula.

$$\begin{matrix} R_2O \\ RO \end{matrix} \cdot R_2O_3 \cdot RO_2$$

This procedure can be illustrated by an example. Using the chemical analysis of a Canadian nepheline syenite (a feldspathic rock) and proceeding according to steps 1 and 2:

Oxide	Weight (lb)		Molecular weight (lb)		Moles
SiO_2	60.7	÷	60.1	=	1.01
Al_2O_3	23.3	÷	101.9	=	0.229
Fe_2O_3	0.08	÷	159.7	=	0.0005
CaO	0.7	÷	56.1	=	0.012
MgO	0.1	÷	40.3	=	0.0025
Na_2O	9.8	÷	62.0	=	0.158
K_2O	4.6	÷	94.2	=	0.049

The weight column is numerically equal to the weight percentage of constituents in the nepheline syenite chemical analysis. The total number of moles of RO plus R_2O oxides (from the last column) is 0.2215. When the number of moles of each oxide is normalized (divided) by this number, the results are:

Oxide	Normalized moles
SiO_2	4.56
Al_2O_3	1.03
Fe_2O_3	nil
CaO	0.05
MgO	0.01
Na_2O	0.71
K_2O	0.22

Arranging in the conventional form results in the empirical formula for this mineral:

$$\begin{matrix} 0.71\ Na_2O \\ 0.22\ K_2O \\ 0.05\ CaO \\ 0.01\ MgO \end{matrix} \cdot 1.03\ Al_2O_3 \cdot 4.56\ SiO_2$$

(9.6)

Such a material does not have a true molecular weight since it is not a compound. It does, however, have a formula weight, computed by multiplying the number of moles of each oxide in the normalized empirical formula by its molecular weight and summing the products. The formula weight for the particular nepheline syenite in the example above is calculated as follows:

Oxide	Moles		Oxide molecular weight (lb)		Weight of oxides (lb)
Na_2O	0.71	×	62.0	=	44.0
K_2O	0.22	×	94.2	=	20.7
CaO	0.05	×	56.1	=	2.8
MgO	0.01	×	40.3	=	0.4
Al_2O_3	1.03	×	101.9	=	104.9
SiO_2	4.56	×	60.1	=	274.0
			Formula weight	=	446.8

If the weight of each oxide were to be expressed as a percentage of the formula weight, then the original chemical analysis would be reproduced.

The batch recipe

It is common practice to specify the composition of glasses, glazes, and enamels by their empirical formulas. To actually produce these ceramics, however, it is necessary to know the specific amount of the various raw materials that must be blended together and heated to produce the final product. The list of the necessary raw materials in the amounts required to produce a desired weight of product is called the *batch recipe.*

To compute the batch recipe, it is first necessary to compute the *formula batch,* which is a list of the amounts of raw materials necessary to produce one formula weight of product. The formula batch is computed from the empirical formula. To clarify the relationship of these terms, the flow diagram shown in Figure 9.1 is helpful. The usual sequence of computations is from left to right. The details of the formula batch will depend on which specific raw materials are used, and the size of the final recipe depends upon the total amount of product to be produced.

The computation of the formula batch from the empirical formula is nothing more than a careful bookkeeping procedure. It consists of the following steps:

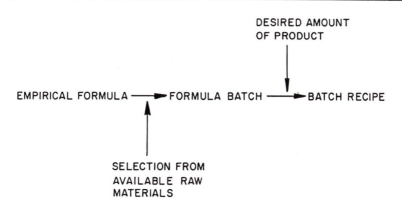

9.1. Relationship between batch calculation quantities.

1. The constituent oxides of the normalized empirical formula are arranged in a horizontal row to form the heads of columns.
2. If alkalies are present, they will usually be added to the formula batch as feldspars or feldsparlike minerals in appropriate amounts to completely satisfy the alkali mole requirements of the empirical formula. Materials such as Na_2O and K_2O cannot practically be added in the form of simple oxides, since such substances would be water soluble and difficult to store. (Na_2O requirements in glasses are usually satisfied with soda ash, Na_2CO_3.)
3. The moles of oxides in the added alkali minerals are subtracted from the empirical formula.
4. Some of the remaining empirical formula requirements can be satisfied by adding raw materials that contribute only one permanent (nonvolatile) oxide. Alumina and silica requirements are left until last.
5. Additional empirical formula requirements may be satisfied by adding raw materials that contribute two oxides.
6. Any remaining alumina and silica requirements are satisfied with materials that contribute only one or both of these two oxides (for example, clay minerals, flint, kyanite, mullite, or corundum).
7. The number of formulas of each raw material added is multiplied by their respective formula weights and summed to yield the formula batch weight.

An example can illustrate this procedure. Let us suppose that the following raw glaze is to be prepared:

$$0.26 \ (Na_2O, K_2O)$$
$$0.44 \ CaO$$
$$0.20 \ ZnO$$
$$0.10 \ MgO$$
$$\bullet \ 0.35 \ Al_2O_3 \bullet 3.00 \ SiO_2$$

For this particular application, the glaze formulator has determined that the relative amounts of Na_2O to K_2O are not especially important as long as their combined total meets the requirements of this empirical formula. (This is not a general rule, but is sometimes acceptable.) From a long list of possible raw materials, the following might be selected:

Nepheline syenite	(Empirical formula in Equation 9.6) Formula weight = 446.8 lb
Whiting	$CaO \bullet CO_2$; formula weight = 100.1 lb
Zinc oxide	ZnO; formula weight = 81.4 lb
Talc	$3MgO \bullet 4SiO_2 \bullet H_2O$; formula weight = 379.3 lb
Clay (kaolinite)	$Al_2O_3 \bullet 2SiO_2 \bullet 2H_2O$; formula weight = 258.1 lb
Flint	SiO_2; formula weight = 60.1 lb

The empirical formula is arranged in row form, and its requirements are satisfied by sequential additions of raw materials in the following manner:

	Na_2O,K_2O	CaO	ZnO	MgO	Al_2O_3	SiO_2
Glaze Formula	0.26	0.44	0.20	0.10	0.35	3.00
0.28 formula						
nepheline syenite	0.26	0.013	...	0.0028	0.29	1.28
Remainder	...	0.427	0.20	0.0972	0.06	1.72

The addition of 0.28 of a formula of the nepheline syenite contributes each oxide in an amount equal to 0.28 times the coefficients in Equation 9.6. The number of formulas that were added to satisfy the Na_2O and K_2O requirement was calculated as being equal to $0.26/(0.71 + 0.22) = 0.28$, where the denominator is the sum of the Na_2O and K_2O in the mineral (Equation 9.6) and 0.26 is the total amount of these oxides required by the glaze formula. In this way, the entire alkali requirement of this glaze is satisfied by 0.28 formulas of the nepheline syenite. The remainder line, obtained by subtraction of the moles of each oxide added from the moles of that oxide required, represents the unsatisfied portion of the glaze formula.

The entire glaze empirical formula can be satisfied in the following way:

Glaze Formula	Na_2O,K_2O 0.26	CaO 0.44	ZnO 0.20	MgO 0.10	Al_2O_3 0.35	SiO_3 3.00
0.28 formula nepheline syenite	0.26	0.013	...	0.0028	0.29	1.28
Remainder	...	0.427	0.20	0.0972	0.06	1.72
0.427 formula whiting	...	0.427	...	...	...	...
Remainder	...	...	0.20	0.0972	0.06	1.72
0.20 formula zinc oxide	...	...	0.20	...	...	...
Remainder	...	...	...	0.0972	0.06	1.72
0.0324 formula talc	...	...	...	0.0972	...	0.13
Remainder	...	...	...	...	0.06	1.59
0.06 formula kaolinite (clay)	...	...	...	...	0.06	0.12
Remainder	...	...	...	...	...	1.47
1.47 formula flint	...	...	...	...	...	1.47
Remainder	...	...	...	...	...	...

Once the glaze empirical formula has been completely satisfied, the number of formulas of each raw material used is multiplied by its formula weight and summed to give the formula batch weight:

Raw material	Number of formulas		Formula weight (lb) (dry material)		Weight added (lb) (dry material)
Nepheline syenite	0.28	×	446.8	=	125.1
Whiting	0.427	×	100.1	=	42.7
Zinc oxide	0.20	×	81.4	=	16.3
Talc	0.0324	×	379.3	=	12.3
Clay	0.06	×	258.1	=	15.5
Flint	1.47	×	60.1	=	88.3
			Formula batch weight	=	300.2

The formula batch weight represents the weight of this particular mixture of dry raw materials that will yield one formula weight of glaze when heated. If a different group of raw materials had been used to formulate this same glaze, then a different formula batch weight would have resulted. The relationship between the formula weight and the formula batch weight is just the theoretical LOI of the batch:

$$\%\text{LOI (raw batch)} = \frac{(\text{formula batch weight} - \text{formula weight}) \times 100}{\text{formula batch weight}} \qquad (9.7)$$

The final step in calculating the batch recipe is to decide how much of each raw material is needed to yield the desired batch size. Suppose for the moment that a 1000-pound batch of raw materials is to be prepared. This constitutes $1000/300.2 = 3.331$ formula batch weights. The weight of each raw material in the formula batch must be multiplied by 3.331. The final recipe for a 1000-pound batch would be:

Nepheline syenite	(125.1) (3.331)	=	416.7 lb
Whiting	(42.7) (3.331)	=	142.2
Zinc oxide	(16.3) (3.331)	=	54.3
Talc	(12.3) (3.331)	=	40.9
Clay	(15.5) (3.331)	=	51.6
Flint	(88.3) (3.331)	=	294.1
	Recipe weight	=	999.8

It should be remembered that this recipe will yield a glaze having the correct empirical formula only if absolutely dry raw materials are used. Raw materials usually contain some moisture which must be accounted for when weighing out a recipe. If the moisture content of each raw material is known, it is an easy matter to correct the recipe. As an example, suppose that the clay has a moisture content of 8% on the undried basis (% moisture [u] = 8%). Then the amount of this moist material that must be added to yield 51.6 pounds of dry material is $51.6/(1 - 0.08) = 56.1$ pounds (see Equation 9.5). Similar corrections may be necessary for other ingredients in the recipe.

Finally, from a practical standpoint the question of how large a recipe to weigh out will usually depend upon the amount of fired product required. Once this has been decided, and after allowance has been made for spillage and other loss, the following simple relationship permits calculation of recipe weight:

$$\text{Weight of recipe} = \frac{\text{raw formula batch weight}}{\text{formula weight of fired product}}$$

$$\times \text{ (weight of fired product to be produced)} \qquad (9.8)$$

Here both the recipe weight and the raw formula batch weight apply to the same specific list of raw materials. This equation can be rewritten to solve for the amount of fired product yielded from any particular weight of raw recipe.

The above formulation procedures are easily adapted to be performed on a computer, and numerous public domain and commercial software packages are available to perform such calculations.

The glaze used as an example here was compounded from raw materials that had not undergone any prior heat treatment. Because of that, the glaze would properly be called a raw glaze. Many times in the preparation of glazes and enamels, a portion of the raw materials is melted together before using it in the recipe. This melting, called *fritting*, produces a glassy material (frit), which is then quenched, ground, and used much like a feldspar additive in the final batch formulation. The reason for fritting a portion of the raw materials is to convert all toxic substances (such as lead oxides) and all water-soluble materials (such as the usual boric oxide [B_2O_3] minerals) into an insoluble glassy form. In addition, any coloring oxides are usually made part of the frit so that they will be uniformly distributed in a glassy form prior to the actual firing of the glaze. A very important parallel to the use of a frit in glazes and enamels is the use of *cullet* (scrap glass) in formulating a glass batch.

Rules for the formulation of frits have been developed by experience over the years. Since most frits used today are produced to order by commercial suppliers, these rules will not be discussed here. Each frit has its own empirical formula and is used just like any other "raw" material to satisfy the formula requirements of the glaze or enamel being prepared. The remaining unfritted raw materials used in the recipe, always including some clay, are wet-milled along with the frit and aid in dispersing the dense frit particles into a stable suspension suitable for application by spraying or dipping.

Numerical factors are available in the literature (for example, Singer and Singer 1964) which allow the calculation of some of the properties of glasses or glazes from the chemical composition. These factors can also be used to formulate glasses having specified properties by selecting alternative chemistries that can be obtained with lower-cost starting materials. The process is somewhat more complex than merely calculating a single physical property for a glass of a known composition. Computer programs are available that can simplify the process. In glaze formulation, the factors are generally used as a guide in raw material selection.

10 Slip

MANY OPERATIONS in ceramic processing involve a slurry of solid particles suspended in a liquid (a slip). These slurry-based operations include separation or blending, spray-drying, wet-milling, filtering, slip-casting, and glaze application. The success of these operations depends not so much on the properties of the suspended solid ceramic material as on the rheological (flow) properties of the slip. This chapter is concerned with measurement of specific gravity and viscosity—the two most important properties of a slip as far as its rheological behavior is concerned.

Specific gravity

The specific gravity (SG) of a slurry is the ratio of the slurry density to the density of pure water at a standard temperature.

$$SG = \frac{\text{density of slurry}}{\text{density of water at } 20°C}$$

A value of density for water of 1.00 g/cm³ (62.4 lb/ft³) is sufficiently accurate for most purposes. If the density of the slip is measured in g/cm³, it will be numerically equal to the specific gravity of the slip.

The specific gravity is always greater than unity for a ceramic-water slip, because it represents an average of the density of the heavy solid particles and the density of water. It follows that the more concentrated the slip, the greater its specific gravity. The concentration of solids in a

properties

slip is often expressed as the mass fraction of solids:

$$\text{Mass fraction of solids} = \frac{\text{weight of solids per unit volume of slurry}}{\text{density of slurry}}$$

The density of a slip can be determined in several ways. If the slip is being freshly prepared, it is a simple task to calculate the density from the amounts of solids and liquid used. If W_S and W_L represent the respective weights of an insoluble powder and a liquid being mixed together, and if ϱ_S and ϱ_L represent the respective densities of the two materials, then

$$\text{Density of slurry} = \frac{W_S + W_L}{(W_S/\varrho_S) + (W_L/\varrho_L)} \tag{10.1}$$

The value used for ϱ_S must usually be the apparent density of the solid particles (see Chapter 8). If the particles are sufficiently small that they do not contain any closed pores, the true density of the solid should be used for ϱ_S.

When it becomes necessary to measure the density of a previously prepared slip, several simple methods are available. A known volume sample of the slip may be collected, using a graduated cylinder or other fixed-volume container, and weighed. This method does not work especially well with highly viscous slips. The simplest measuring method, and one which can be employed directly in a blunger, utilizes a hydrometer— a closed, weighted tube of fixed displacement that will partially submerge in a fluid (Fig. 10.1). The lower the density of the slip, the deeper the level of submergence of the hydrometer. The depth of submergence is read on the stem of the hydrometer, which is usually graduated directly in specific gravity units. A hydrometer can also be used for measuring specific gravity of solutions, for example, for checking antifreeze concentrations (and therefore freezing points) in automobile cooling systems.

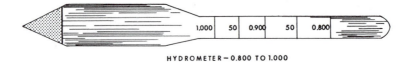

HYDROMETER – 0.800 TO 1.000

10.1. Hydrometer for specific gravity measurements. This particular instrument is intended for organic liquids with densities less than water.

It is frequently necessary to adjust the specific gravity of a slip by adding water or by blending together two slips of differing density. The two liquids may be proportioned by weight or by volume. If W, V, and ϱ represent the weight, volume, and density of liquids 1 and 2, the density resulting from blending these liquids is given by

$$\varrho_{1+2} = \frac{W_1 + W_2}{(W_1/\varrho_1) + (W_2/\varrho_2)} \tag{10.2}$$

or by

$$\varrho_{1+2} = \frac{V_1\varrho_1 + V_2\varrho_2}{(V_1 + V_2)} \tag{10.3}$$

Viscosity

All fluids exhibit some resistance to being stirred or sheared. This property is termed the viscosity of the fluid and is a very important quantity to measure and control in ceramic slip processing. The usual units of viscosity are centipoise (1 centipoise = 0.01 g/cm-second), and a rough rule of thumb is that pure water possesses a viscosity of 1 centipoise at room temperature. The usual symbol for viscosity is μ.

The viscosity of a fluid expresses the relationship between the unit force (stress) being applied to cause a fluid to shear (flow) and the rate of shear that the fluid exhibits. Rheologists (scientists who study the flow characteristics of fluids) often graph shearing rate versus shearing stress for the fluid. When the resulting consistency curve is linear, as in Figure 10.2, the fluid is called a *Newtonian* fluid, and the reciprocal of the slope is equal to the viscosity.

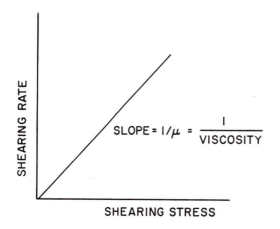

$$\text{SLOPE} = 1/\mu = \frac{1}{\text{VISCOSITY}}$$

SHEARING STRESS

10.2. Consistency curve for a Newtonian fluid.

Because the viscosity is a measure of the resistance of a fluid to flowing, it is a property that has an important influence on pumping, filtering, pouring, mixing, spray-drying, and settling of ceramic slips. Since gases also possess characteristic viscosities (much lower than for liquids), viscosity is also of great importance in drying, combustion, pneumatic conveying, and dust collecting in a ceramic plant.

The measurement of liquid viscosities can be performed by a variety of methods. The so-called single-point methods are certainly the simplest and least expensive. These methods measure the shearing rate at one shearing stress, to yield a single point on the consistency curve. If the fluid is Newtonian, a straight line drawn through the origin and this single point determines the viscosity. A falling ball viscometer may be employed with Stokes's law (Equation 11.1) applied to the observed settling rate of a dense sphere in the liquid. The capillary tube or Poiseuille viscometer (Fig. 10.3) is often used with ceramic slurries. Here the viscosity is determined from the rate of flow of the fluid through a capillary of known size. Various orifice viscometers (such as the Saybolt viscometer) determine viscosity from the time a given amount of fluid takes to drain through an orifice in the bottom of a standardized cup or cylinder.

Multiple-point methods for measuring viscosities employ a small paddle or solid bob rotating in a stationary cup filled with liquid (or vice

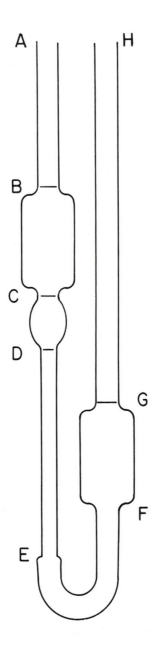

10.3. Diagram of a Poiseuille viscometer. Liquid is introduced through H until it reaches level G. Suction is applied at A to draw the liquid above mark B. The liquid is then allowed to flow under its own head through capillary D-E, and the time required for the liquid level to drop from B to C is a measure of the viscosity.

versa) and are capable of determining a number of different points on a consistency curve. Two such rotational viscometers are shown in Figure 10.4. The instrument in Figure 10.4(a) employs a variable-speed rotating cup, with the shearing stress measured by determining the angle of twist of an anchored stiff wire supporting the suspended bob. Figure 10.4(b) shows a hand-held instrument using a variable-speed rotating bob to be inserted in a stationary container of liquid. The apparent viscosity at each shearing rate is indicated on a dial linked to the bob by means of a torsion spring. Recording viscometers are also popular, yielding much information relating to shearing stress, shearing rate, and viscosity, and are capable of producing entire consistency curves (Fig. 10.4[c]).

These instruments are most important for measuring non-Newtonian fluids. Most ceramic slips exhibit nonlinear consistency curves, the two most important types of which are shown in Figure 10.5. Curve *a* represents a *dilatant* fluid, which is one that becomes stiffer at high shear rates than at low shear rates. Curve *b* is more commonly encountered in ceramic slips. This represents a *pseudoplastic* fluid, which becomes less

10.4. Rotational viscometers:

(a) MacMichael viscosimeter (*courtesy Fisher Scientific Co.*).

(b) Brookfield viscometer (*courtesy Brookfield Engineering Laboratories, Inc.*).

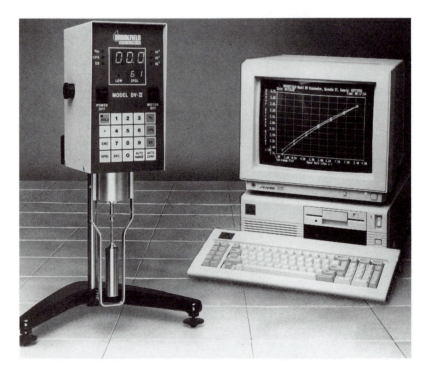

(c) Brookfield DV-II viscometer with computer acquisition and display of data (*courtesy Brookfield Engineering Laboratories, Inc.*).

stiff at high shear rates than at low rates. If a pseudoplastic liquid slowly regains it stiffness after shearing is stopped, it is said to show *thixotropic* behavior. Non-Newtonian fluids cannot be characterized by a single viscosity and thus single-point measurement methods offer only limited information. Rheology is a complex subject, and readers are encouraged to investigate texts such as Moore (1965).

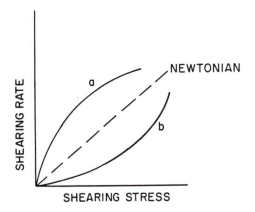

10.5. Consistency curves for non-Newtonian fluids. Curve *a* exhibits dilatant behavior and curve *b* exhibits pseudoplastic behavior.

Viscosities of slips depend strongly on temperature, specific gravity, the presence of deflocculants, and the particle size distribution of the materials in the slurry.

Casting tests

An important property of a ceramic casting slip (see Chapter 4) is the rate of buildup of the cast wall in the mold. As soon as the slip enters the plaster mold, the wall begins to form rapidly. As the wall thickens, the water in the slip has more and more difficulty in passing out to the plaster through the cast wall, and the buildup rate drops off. It is often found that the wall thickness T is related to the casting time t by

$$T = C\sqrt{t} \tag{10.4}$$

where C is a constant that must be determined experimentally. The value of C can be increased by raising the specific gravity or lowering the viscosity of the slip. After a casting slip has been prepared, it is common practice to make trial casts in small plaster molds, and to graph thickness versus casting time. This relationship can then be used to determine the casting time needed for any desired wall thickness. A casting test also serves to point out draining problems, which can lead to an unsatisfactory surface finish on the formed piece.

Many traditional and advanced ceramic products involve processes utilizing slurries at some point, especially due to the increased use of spray driers, either for the final body processing step or as an intermediate step. The combined requirements of mixing, pumping, filtering, and casting of slips make a knowledge of slurry properties more important now than at any time in the past.

11

Measurement
and powder

THE PROCESSING OF CERAMIC raw materials could well
be called fine particle technology. Throughout the prefiring
manufacturing stages, considerable effort is expended in
producing fine particles, separating them according to size,
then often blending the fractions together to achieve proper
size proportions for good packing during forming. None of
these particular operations can be carried out successfully
unless procedures are available for measuring the size of the
particles being handled. The science and technology of particle size measurement has produced hundreds of techniques
for particle size determination. This chapter considers the
general aspects of several of these procedures in common
use in the ceramic industries.

Particle size distribution

A powder is seldom encountered that contains particles of only one
size, although such "monodisperse" powders are of considerable interest
for production of certain advanced ceramics. Powders normally contain
particles in a range of sizes, and it is more proper to speak of the particle
size distribution of the powder rather than its particle size. The distribution is a description of the relative amounts of particles of various sizes
included in the powder. The relative amounts are most often expressed in
terms of mass fraction, the fraction of the weight of the entire powder
that constitutes particles of a particular size. These relative amounts can
be tabulated versus particle size, but it is a common practice to graph

# of particle size
surface area

this information, since this allows a more rapid visualization of the particle size distribution than does tabulation. Figure 11.1 shows the two most popular ways of graphing particle size distributions. Figure 11.1(a) shows the mass fraction of each size plotted versus size. The bell-shaped curve is typical of most powders, with the commonest particle size in the powder being the one directly under the maximum on the curve. Figure 11.1(b) shows the cumulative percentage finer than a certain size plotted versus size. This graph has the advantage that any size can be selected and the total amount of the powder finer (or coarser) than that size can be read directly. It is important to remember that the quantity plotted is mass fraction (or cumulative mass fraction) and not number count or percentage of particles of each size. Because individual large particles weigh more than individual finer particles, equal mass fractions of two different-sized particles do not consist of equal numbers of particles of the two sizes. Certain techniques are capable of producing number versus size distributions, and these have certain advantages over the commoner mass fraction versus size plots.

11.1. Graphic presentation of particle size data:

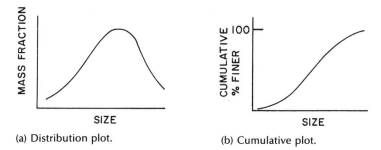

(a) Distribution plot.　　　　(b) Cumulative plot.

The methods of particle size measurement described in this chapter require that the total amount of material actually analyzed be small. A means must therefore be available for removing a small sample from a large process stream before an analysis can be made. Since results of the analysis of the sample are useful only if they are descriptive of the entire process stream, an improperly selected sample may well result in incorrect and costly processing decisions for the whole stream. This point cannot be emphasized too strongly. Each plant must work out its own particular sampling scheme to assure a successful operation.

Testing sieves

The standard procedure for determining particle size distribution of powders coarser than about 50 μm is the use of testing sieves. These sieves are a series of carefully made woven wire screens having openings of standardized sizes. The sieves are usually made in 8-inch diameters with brass sides configured to allow a group of sieves to be stacked or "nested" together, with the coarsest sieve at the top and the finest at the bottom (Fig. 11.2). Although the sieve-opening sizes are accurately known and standardized, it is common practice to refer to sieves by their mesh number, which is the number of sieve openings per lineal inch of sieve surface. The higher the mesh number, the smaller the sieve opening. The relationships between opening size and mesh number for the two most common series of testing sieves are given in Table 11.1.

TABLE 11.1. **Standard testing sieve apertures**

ASTM Sieve No.	ASTM Mesh Opening (*mm*)	Nearest Tyler Sieve Mesh No.	Tyler Mesh Opening (*mm*)
3	6.35	3	6.68
4	4.76	4	4.70
6	3.36	6	3.33
8	2.38	8	2.36
12	1.68	10	1.65
16	1.19	14	1.17
20	0.84	20	0.83
30	0.59	28	0.59
40	0.42	35	0.42
50	0.297	48	0.295
70	0.149	100	0.147
100	0.105	150	0.104
200	0.074	200	0.074

Sources: American Society of Testing and Materials, ASTM E 11, in *Annual Book of ASTM Standards, 1970;* and *Testing Sieves and Their Uses,* Mentor, Ohio, W. S. Tyler, Inc., Handbook 53, 1967.

11.2. Set of nested testing sieves (*courtesy W.S. Tyler Inc.*).

If a weighed sample is introduced onto the top sieve in a nested stack and the stack is shaken, the sample will sift down and separate into sized fractions on the progressively finer sieves. Material that passes through the finest sieve is caught in a pan at the bottom of the stack. Machines are available to shake the stack for a preset time using a reproducible motion (Fig. 11.3). When sieving is complete, the sieves can be unnested, and the weight of material retained on each sieve and in the pan recorded. These data can then be tabulated versus sieve-opening size or plotted as shown in Figure 11.1. When plotting, the average diameter of particles in a given fraction is taken as the average aperture size of the two screens involved; for example, with Tyler sieves, -10 mesh $+ 14$ mesh material, has an average diameter of $(1.65 + 1.17)/2 = 1.395$ mm (0.0549 inch).

11.3. Ro-tap testing sieve shaker (*courtesy W.S. Tyler Inc.*).

A typical sieve analysis of a material might be reported as:

+ 10	10%
−10 + 14	20
−14 + 20	35
−20 + 28	20
−28 (pan)	15
	100%

This tabulation indicates that 10% of the sample weight did not pass through a 10-mesh sieve and thus must be coarser than 1.65 mm (0.065 inch). Also, 20% of the sample passed through a 10-mesh sieve but did not pass through a 14-mesh sieve (sometimes expressed as 10/14 rather than −10 + 14); this part of the sample must include particles between 1.65 mm (0.065 inch) and 1.17 mm (0.046 inch). A total of 30% (10% + 20%) of the sample is coarser than 14 mesh (1.17 mm, 0.046 inch). Similar statements can be made concerning the −14 + 20 and the −20 + 28 fractions. Finally, 15% of the sample passed through a 28-mesh sieve and must therefore be finer than 0.59 mm (0.0232 inches). Nothing can be said about how much finer than 28 mesh this material is unless additional sieves are employed, and nothing more can be said about the coarseness of the +10 fraction.

Sieve analyses can be performed dry or on slurries. The wet procedures are commonly used for very fine powders where water helps to carry the fine particles down through the stack.

A number of ASTM standard procedures are available for sieve analysis of ceramic materials, spelling out such factors as sample size, sample selection, and sieving time. Several of these procedures are listed in Table 11.2.

TABLE 11.2. **ASTM standard procedures for sieve analysis of ceramic materials**

Test Designation	Applicable Ceramics
C 92–88	Refractory materials
C 429–82 (1987)	Materials for glass manufacture
C 322–82 (1988)	Sampling of whiteware clays
C 371–89	Nonplastic whiteware materials
C 325–81 (1988)	Whiteware clays (wet analysis)
E 11–87	Wire cloth sieve specifications

Woven-wire sieves, especially of the finer opening sizes, are very delicate and are subject to serious damage if mistreated. The wire cloth itself can easily be torn or pierced during cleaning, and even slight pressures on the wire cloth surface may distort the sieve openings. Either type of damage is permanent, and a damaged sieve must be discarded. Cleaning is normally done by gently brushing the back side of the sieve cloth with special brushes purchased from the sieve manufacturers. Washing with soap and warm water is permissible, but strong reagents must be avoided.

Sedimentation methods

Stokes's law describes the steady state settling velocity of spherical particles in a continuous still fluid medium. This law may be written as

$$v = \frac{(\varrho_s - \varrho_f)gD^2}{18\mu} \qquad (11.1)$$

where v is the steady state velocity of settling of a particle of diameter D and density ϱ_s in a fluid of density ϱ_f and viscosity μ; g is the gravitational constant. This expression can be transposed and solved for diameter

$$D = 0.136 \left(\frac{\mu v}{\varrho_s - \varrho_f}\right)^{1/2} \qquad (11.2)$$

This expression yields particle diameters in centimeters if the viscosity is expressed in centipoise, the observed settling velocity is expressed in cm/second, and the densities are expressed in g/cm^3. Using this relationship, it is possible to set up experiments where a sample of particles is allowed to settle in a fluid free of convection currents, with the observed settling velocities being converted to particle sizes.

Several complications need to be considered prior to starting a sedimentation experiment. First, the fluid in the settling chamber must be free of convection currents that would disturb the normal free settling of the particles. These extraneous currents can be eliminated only if the settling chamber is uniform in temperature throughout. Second, Stokes's law is strictly valid only for spherical particles. Since most particles of interest are not spherical, the size calculated from the settling velocity will not be an exact diameter, but rather an *equivalent spherical diameter*. If the true diameter is needed, shape corrections must be applied. Third, the presence of other particles interferes with the free settling of each particle, especially if the particles tend to flocculate (agglomerate). This problem can be overcome by using very dilute suspensions for sedimentation experiments, and also by using dispersing aids (deflocculants) to avoid agglomeration.

Finally, real powders consist of a mixture of many particle sizes. Stokes's law is valid for each particle size, so that every particle of the same size settles at the same rate, but the presence of many different sizes means that many different settling velocities are occurring simultaneously. The key to unraveling this tangled group of simultaneous velocities lies in the fact that at a fixed depth within the initially uniform suspension there are discrete settling time intervals after which no particle larger than a certain size is ever found. The larger particles settle more rapidly than the smaller particles, so that at the fixed depth, the average particle size present becomes finer and finer as time goes by. As this progressive settling occurs, the particles reaching the bottom of the settling chamber form a sediment that is initially mostly coarse, but which collects ever-finer particles with time. The problem of separating this complicated process into individual settling rates (and hence individual particle sizes) is thus reduced to one of continuously sampling the process either at a fixed depth in the fluid or by catching the sediment as it builds up. Despite all of these complicating factors, sedimentation methods of particle size determination are standard procedures of the ceramic industries when material finer than about 50 μm is involved. They are particularly important to those branches of the industry using clays and fine oxides.

Traditionally, the most common sedimentation procedure involves the use of a hydrometer (see Fig. 10.1) to periodically sample the changes in density of the suspension at a known depth, a method spelled out in detail in ASTM Standard Method of Test D 422–63 (1990). A second method, not as widely accepted, employs a special pipette to remove small samples of the suspension periodically from a fixed depth for drying and weighing. Both procedures, which may involve the use of a centrifuge to shorten settling time (by increasing the value of g in Equation 11.1), yield particle size distributions.

The sediment buildup in a sedimentation experiment can be followed by using a balance pan suspended at a fixed depth in the settling column to collect all particles reaching it. This arrangement is usually automated to continuously record weight of material on the pan as a function of time. These data will yield a particle size distribution. The sedimentation fluid employed in such a device may be a gas rather than a liquid to speed up settling rates and shorten the time needed to complete a measurement.

The SediGraph particle size analyzer (Fig. 11.4) measures the sedimentation rates of slurries and automatically plots the mass percent distribution in terms of equivalent (Stokes's) spherical diameter. A finely collimated X-ray beam is used to determine the concentration of particles remaining at decreasing sedimentation depths as a function of time. The instrument typically yields a particle size distribution from 50 μm down to 0.18 μm. This instrument has become quite common in ceramic laboratories.

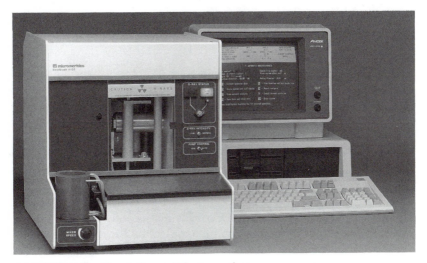

11.4. Sedigraph sedimentation particle size analyzer (*courtesy Micromeritics*).

Light-scattering methods

The MICROTRAC particle size analyzer (Fig. 11.5) is typical of a number of instruments that use the scattering of a laser light beam by particles in a flowing liquid stream to determine certain parameters of particulate distribution. Fraunhofer diffraction, using low-angle forward light scattering, is used for the particle size range of 2 μm to 2000 μm. Other scattering techniques can be used for determining the particle size distribution down to 0.1 μm. These techniques lead to number-based rather than weight-based distribution information.

11.5. MICROTRAC II light-scattering particle size analyzer
(*courtesy Leeds & Northrup*).

Microscopic methods

The optical microscope (described in Chapter 12) can be used to measure particle size and shape directly for particles ranging from about 100 μm down to about 1 μm in diameter. The electron microscope can be used for finer particles, but its use is not described here. The particles may be either dispersed or in a dense fired mass, with the details of measurement being somewhat different for each. The exact magnification at which the particles are viewed must be known before any size determination can be made. The total nominal magnification, calculated as the product of the nominal magnifications of the objective and ocular systems, is usually not accurate enough, and an object of known size must be used to calibrate the exact magnification. Special care is necessary in preparing a representative sample, since it is impossible to perform an optical analysis on a sample weighing more than a few tenths of a milligram. A 0.1-mg sample of ceramic particles consisting of cubes 5

μm on a side would contain about 320,000 individual particles—far too many to be counted conveniently. Selecting a fractional milligram sample to truly represent a tonnage process stream is an exacting task indeed. In any case, the field of view of the microscope is so limited that a number of different locations within the sample must be observed to offer a truly representative group of particles.

For analyzing the size distribution of unconsolidated particles, the sample is usually dispersed in a liquid having an index of refraction very different from that of the particles. A drop of this dispersion is placed on a glass slide and viewed in transmitted light.

Measurement of the particles may be made in place with a filar ocular. This device replaces the normal microscope ocular and contains a hairline that can be moved across the field of view by means of a micrometer knob. The hairline is aligned with one edge of a particle, and the micrometer is read. The hairline is then swept across the particle to the opposite edge, and the micrometer is again read. The difference in micrometer readings multiplied by a magnification factor yields the particle dimension. A second means of measurement in place uses a special insert in the microscope called a reticle, which superimposes a group of numbered circles or squares of different sizes on the field of view of the microscope. The particle is judged to be comparable in size to one of the numbered figures, and with a suitable magnification correction, a particle size is determined.

It is often more convenient to measure the size of uncompacted particles by first taking a picture of them through the microscope (photomicrograph), then taking measurements from the photograph; this technique is always used with an electron microscope. ASTM standard E 20–85 describes in detail a method for preparing samples and making particle size measurements from photomicrographs. Either the projected area of the particle or a linear measurement can be determined from the photograph, and a particle size can thus be determined.

If the particle size of a consolidated ceramic material (called its grain size) is to be determined, the sample must first be prepared for observation using the mounting and polishing procedures described in Chapter 12. Either transmitted or reflected light may be used for viewing if the sample is transparent, but the normal viewing mode for grain size measurement is reflected light. Comparison reticles can be used for measurements in place, but normally a photomicrograph is taken. The photomicrograph can be compared to standard grain-size charts (for example, ASTM E 112–88) and an average grain size can thus be assigned. A whole set of size, area, and volume parameters can also be determined for the sample by counting the number of grain boundaries crossing random lines drawn on the photomicrograph (see, for example, Fulrath and Pask 1968).

Automated image analyzing systems (Fig. 11.6) now combine microscopic techniques with computer analysis, yielding not only particle size data, but also data on the distribution of mineralogical phases and pore volume (based on differences in reflectance). This removes much of the tedious work of analyzing visual images by manual techniques.

11.6. Computer-based image analysis system, which provides quantification of images based on their gray level differences and morphology. These include particle size, shape, and distribution; area and volume fractions; and length, width, and other dimensional characteristics (*courtesy LECO Corporation*).

Specific surface area

The specific surface area (the total surface area per unit weight) of ceramic powders is an important parameter related to particle size, but the measurement of this property can be very complex for a complete analysis. For that reason, the property is usually not measured for daily quality control purposes in most ceramic laboratories. Suppliers of ceramic powders often report specific surface area data for their materials, but even then, the daily control parameter usually is the particle size distribution rather than the specific surface area. Instruments usually determine specific surface area by measuring the quantity of gas adsorbed onto the surface of a powder by sensing the change in the thermal conductivity of a flowing mixture of an adsorbate (often N_2) and inert carrier gas (often He). Multiple-point so-called BET adsorption instruments can yield information about the distribution of open porosity as well as specific surface area (Fig. 11.7).

11.7. AUTOSORB-6 fully automated single- or multiple-point gas adsorption surface area and pore size analyzer (*courtesy Quantachrome*).

12

Structure

THE MICROSCOPIC EXAMINATION of individual ceramic grains and fired consolidated ceramic structures mentioned in Chapter 11 is discussed in detail in this chapter. Emphasis is placed on the preparation of specimens for examination with an optical microscope, which utilizes either transmitted or reflected visible light. Since the electron microscope and the scanning electron microscope (SEM) are also of importance in examining the structure of ceramics, they are described briefly. Finally, X-ray diffraction and several chemical analysis techniques are discussed.

Microscopic analysis

The unaided human eye can resolve two adjacent particles if the diameter of the particles is greater than about 0.2 mm. Smaller particles must be observed using microscopic techniques.

The amount of useful magnification obtainable using a light microscope may vary from as low as 5 times (5×) to as high as 1000 times (1000×). The light microscope has a limit of resolution that is governed by the wavelength of visible light. If objects smaller than about 1 μm (1 micrometer, or slightly less than 0.000040 inch) are to be viewed, an electron microscope must be used. The limit of resolution of an electron microscope is on the order of 0.001 μm, and magnifications of 1,000,000× can be achieved.

Optical microscopes can be arranged to view an object by either reflected or transmitted light. If the object is either fairly thick or opaque, the image must be formed by reflecting light off the surface of

and composition

the object and into the imaging lenses. If the object is transparent and is sufficiently thin, an image can also be formed from light transmitted through the object from the back side. For general ceramic work, a microscope capable of both kinds of operation is generally required.

The microscope

All microscopes possess certain basic features, among which are a source of illumination (the illuminator) and a set of lenses for concentrating this illumination onto the sample (the condenser); a platform on which the sample is supported (the stage); a set of lenses that form a magnified image of the sample inside the body of the microscope (the objectives); a set of lenses to further magnify the image and project it into the eye or a camera (the ocular or eyepiece); and a means of focusing the image either by moving the objective or the stage so that their separation distance can be changed.

Figure 12.1 shows two variations of the basic optical microscope. They are double ocular (binocular) microscopes, but many instruments use a single eyepiece (monocular) design. Four objectives (Fig. 12.1[a]) and three objectives (Fig. 12.1[b]) of differing magnifications mounted on a revolving turret are shown, but some microscopes can accommodate only one objective at a time. The total magnification of the microscope image is equal to the product of the magnification of the objective and the ocular (for example, $8\times$ ocular with $32\times$ objective gives total magnification of $256\times$).

The illuminator indicated in Figure 12.1(a) is called a vertical illuminator and is used only for reflected light observations. The condensers for vertical illumination are actually built into the objectives. The mirror below the stage is used to reflect light from an external illuminator up through the sample for transmitted light observation. The stage shown

213

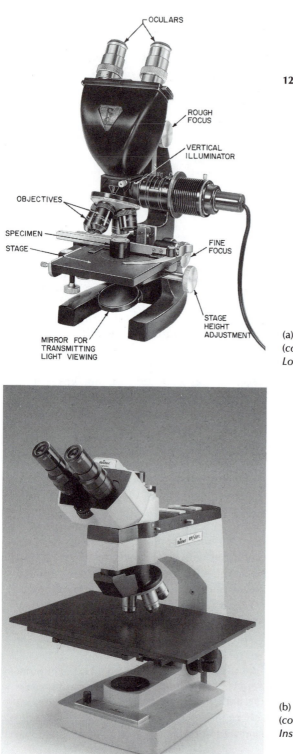

12.1. Optical microscopes:

Labels on figure (a):
OCULARS
ROUGH FOCUS
VERTICAL ILLUMINATOR
OBJECTIVES
SPECIMEN
STAGE
FINE FOCUS
STAGE HEIGHT ADJUSTMENT
MIRROR FOR TRANSMITTING LIGHT VIEWING

(a) Binocular microscope (*courtesy Bausch and Lomb*).

(b) Binocular microscope (*courtesy Cambridge Instruments*).

can be moved horizontally in two mutually perpendicular directions by controls below the stage, but many microscopes have stages that also rotate around the center line of the instrument. Rough and fine focusing of the microscope shown is accomplished by moving the stage up and down while the objective remains fixed.

The objectives are the heart of any microscope, because they are the complex lens system that forms the image. They are usually ground and assembled with great care to eliminate various kinds of optical aberrations which would detract from the quality of the image. They are very expensive and must be treated with utmost care. A set of objectives for a microscope usually includes three to five separate objectives ranging in magnification from about $5 \times$ up to about $100 \times$. Each objective has a characteristic working distance – the distance from the lens to the object needed to achieve focus. In general, high-magnification objectives have short working distances, which means that they will almost come into contact with the object when they are focused (actual contact with the object may damage the lens). Occasionally, an oil-immersion objective is used, where a drop of special oil provides an "optical link" between a high-power objective and the sample surface. In addition to short working distance, a high-magnification objective will also have a very limited depth of field – that is, minute changes in the focus control will cause the image to go out of focus. For this reason, it is always easier to first focus with a low-power objective and then switch to a high-power objective, provided that the microscope has a multiobjective turret.

In addition to the basic features already described, many microscopes have additional accessories. Provisions are often made for insertion of various illumination apertures and field diaphragms as well as polarizer-analyzer pairs. Quite often, eyepieces with special inscribed patterns, called reticles, are used to superimpose these patterns on the image for measurement purposes. Some means is usually provided for attaching a camera to the microscope in order to take a photograph of the image (a photomicrograph). Details of the use of these refinements will not be given here.

A particular kind of microscope that usually incorporates all of the features described above is the metallograph. Figure 12.2 shows two metallographs and illustrates the unusual arrangement of the lenses in such instruments. The objectives are usually mounted to point upward through a hole in the stage. The sample is placed face down on the stage over the hole. Such instruments are normally used only for reflected light observation with the illuminator commonly being a halogen high-intensity lamp.

All lenses in a microscope are susceptible to serious damage from

12.2. Metallographs (*courtesy LECO Corporation*):

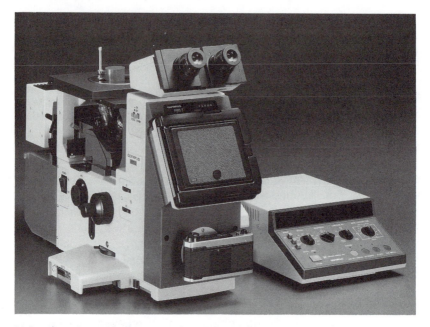

(a) Benchtop inverted-stage research metallograph with bright- and dark-field, polarized light, interference contrast, zoom and full photomicrography capabilities.

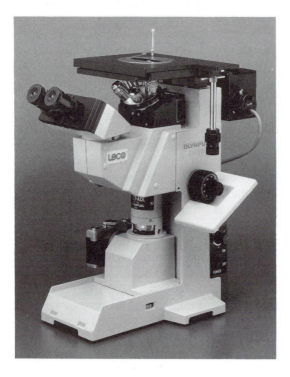

(b) Benchtop inverted-stage metallograph for routine microscopy and photomicrography with reflected light.

even relatively minor mishandling. The lenses are generally coated with a soft antireflecting material that is readily scratched. The lenses must never be allowed to touch any solid material and especially not the abrasive ceramic sample being viewed. Careful handling of the lenses prevents them from being damaged by fingerprints. If it becomes necessary to clean the lenses, only special lens paper is used along with the cleaning solvent recommended by the lens maker. Ordinary tissues will scratch the coatings on the lenses. Irreparable harm will be done to an objective if any etchants used in sample preparation are not thoroughly removed before placing the sample on the microscope stage. To keep the microscope clean and operating properly, it should always be kept in a dry room away from chemical reagents and should be kept boxed or otherwise covered when not in use.

Examination of individual ceramic grains

The microscopic examination of individual uncompacted ceramic grains is usually performed to determine their size, shape, or mineralogy and is usually made with transmitted light. Such tests are often performed on raw materials when they are received from the supplier, but occasionally size and shape are determined on material just prior to forming. The grains to be examined are suspended in a drop of special oil on a glass slide under a cover glass.

A number of methods exist for measuring grain size with a microscope as described in Chapter 11. Grain shape may be observed directly, but irregularly shaped grains usually lie on the glass side in such a way that the side having the greatest area is flat on the slide. The smallest dimension of the grain may thus be missed, which can lead to a mistaken idea about the shape of the grain.

The mineralogical identification of ceramic grains is a complex procedure. Sometimes the color of the grain can be helpful, and the presence of well-defined cleavage planes is also an identification aid. The most important way of identifying a mineral with a microscope, however, is by observing the way it interacts with light. Every mineral has a characteristic index of refraction, which is a measure of the velocity of light in that material. Whenever a light beam traveling through a suspending oil having one index of refraction passes into a crystal having a different index, the light beam bends and the boundary between the oil and the crystal will be visible. If both materials have the same index, the boundary is not seen, and the crystal is "invisible." A great many oils of known index of refraction are available, and these can be used to suspend the grains for microscopic observation. In the simplest case when an oil having the same index as the grains is used, the grains will not be seen in the microscope field, and their index of refraction is thus determined. Tables giving indices of refraction of minerals can then be used to help identify the grain. A mineral which is anisotropic (properties vary with direction) has more than one index of refraction, according to the orientation of the crystal.

Finally, certain kinds of minerals having certain kinds of crystal structures show interference patterns when viewed in the microscope under polarized light. The shapes of these patterns and the way in which they change when the microscope stage is rotated can also be of considerable help in identifying the mineral. The reason for these interference patterns and their interpretation is beyond the scope of this book. Details on optical procedures for identifying minerals will be found in optical mineralogy textbooks.

Examination of consolidated specimens with transmitted light

The major portion of the work involved in making microscopic observations on consolidated ceramic specimens (sometimes called ceramography) is concerned with the preparation of the specimen for viewing. Such preparation usually involves cutting a representative specimen from a larger block of material, mounting in a special holder, grinding to produce a flat surface, and polishing the surface smooth. Since many ceramic materials are very hard, several of these operations can involve considerable effort unless diamond tooling is utilized. Often, particularly if the specimen material is quite porous, it must first be impregnated with a resin to give it enough strength to remain intact during grinding.

When the specimen is to be examined with transmitted light, it must be thinned to roughly 30 μm thickness before polishing. The specimen is cut to as thin a slab as is convenient with a diamond saw. The specimen can then be ground flat and cemented onto a glass slide with a resin. The slide can be clamped into any one of several available thin-section holders. Grinding to the required thickness is done wet and can be accomplished by using loose abrasive grains on either a glass plate or a rotating metal wheel, or by using abrasive papers (Fig. 12.3). Abrasive papers and powders are available in a number of grit sizes. The coarser abrasives remove material more rapidly than do the finer abrasives. When changing to a finer abrasive, the specimen must be thoroughly washed to prevent carryover of coarse abrasive. For very hard materials, grinding is done with a bonded diamond wheel. All grinding operations are done wet.

After grinding, the thin section must be polished to remove the heavily scratched layer introduced during grinding. Polishing is usually done on a wax- or cloth-covered wheel using a paste containing very fine

12.3. Grinding and polishing apparatus for ceramography
(*courtesy LECO Corporation*):

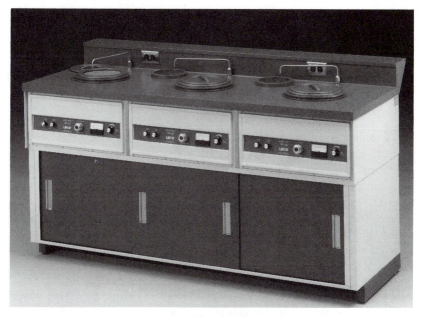

(a) Triple-wheel console for manual wet-grinding and polishing.

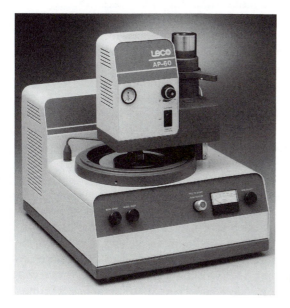

(b) Automatic grinder and polisher.

diamond or alumina powder. It requires several steps using successively finer powders and finishing with a powder having a nominal size of about 0.05 μm. The sample should be cleaned in an ultrasonic cleaner between each polishing step to prevent the mixing of coarse and fine abrasives. As with grinding, the polishing operation is always done wet.

Once the final polishing operation is completed, the slide with the thin section cemented onto it can be removed from the holder and placed on the microscope stage for observation. The specimen is illuminated from the back side, usually with polarized light.

Examination of consolidated specimens with reflected light

Most routine examination of ceramic materials is done on opaque specimens using light reflected from a polished surface. The preparation of a specimen for examination involves several steps in common with thin-section preparation, but the thinning step is eliminated and a final etching step may be added. After the specimen has been cut from the parent block of material, it is embedded in a plastic holder (mount) approximately 1 inch in diameter by perhaps ¾ inch thick. The embedding procedure is usually carried out under pressure in a heated steel die starting with the plastic in granular form. Often it is necessary to impregnate a very porous specimen with a resin prior to mounting so that it will have sufficient strength to survive the mounting and subsequent steps. If the specimen is extremely fragile, it can be mounted by using a liquid resin that cold-hardens in a mold without applying pressure.

Once the mounting procedure is completed, the surface to be viewed is ground flat using procedures similar to those used in grinding down thin sections. The specimen mount is sometimes held by hand for this operation, but commercial equipment is available that will hold several mounts at the same time in contact with a grinding wheel.

After the surface has been ground flat, it must be wet-polished on a cloth-covered surface with several successively smaller sizes of either diamond or alumina powder until all scratches are removed. This is usually accomplished using a wheel, but commercial equipment is also available that moves the mounted specimen freely across a polishing cloth by means of a sixty-cycle vibratory motion.

The polished specimen will have a mirror finish, but viewing it with light reflected off this surface may show little detail. A final etching step may be required to reveal the various features of the microstructure. When the ceramic surface is contacted with an etchant, less chemically resistant parts (usually adjacent to grain boundaries) dissolve more rapidly than other parts, leaving a selectively roughened surface for viewing in the microscope. Many etchants can be used for ceramics, the most popular being various mixtures of hydrofluoric, nitric, hydrochloric, sulfuric, and phosphoric acids. The etchant is usually warmed to increase its rate of solution. (**Warning:** While all these reagents are capable of causing serious burns, hydrofluoric acid is a particularly dangerous material. Unusual precautions are necessary when using this acid.)

After etching, neutralizing, and washing, the specimen mount is placed on the microscope stage for viewing. Since objectives will be in close proximity to the specimen during viewing, serious damage to the lenses will result unless all traces of etchant have been removed. For most acid etchants, a concentrated ammonium hydroxide soak followed by several water washings is usually sufficient. If a hydrofluoric acid etchant has been used, however, a normal basic solution is not adequate to neutralize this material completely, and etching of the objective lens surface may result. The only truly effective neutralization procedure for hydrofluoric acid etchants is an hour-long soak in a concentrated aqueous solution of ammonium pentaborate ($NH_4B_5O_8 \cdot 4H_2O$). This should be followed by several water washings. Distilled water is always used for washing, and drying should be done in such a way as to prevent formation of water spots.

Electron microscopes

The limit of resolution of an ordinary light microscope is about 1 μm, and if objects or microstructural features smaller than this are to be viewed, an electron microscope must be used. The traditional electron microscope uses magnetic lenses to produce an image from a narrow beam of high-energy electrons transmitted through the very thin specimen; for this reason, this kind of microscope is called a transmission electron microscope (TEM). The specimen is placed inside the microscope column, which must be evacuated before operation. The image is either viewed on a fluorescent screen or projected onto a photographic plate for later development and enlargement.

The maximum specimen size that can be accommodated in a transmission electron microscope is about 1/8 inch in diameter, but the maximum thickness than can be utilized is on the order of a few tenths of a micrometer. It is difficult to thin a specimen down this much, and for that reason, a very thin plastic or carbon replica of the surface is sometimes produced for viewing. The replica is produced by vapor deposition as a coating on the specimen surface which can be stripped off and mounted on a rigid metal grid before being placed inside the microscope. Tiny individual grains can be viewed directly by placing them on a supporting grid. Photographs of these grains can be used directly to measure particle size. Specimens are often shadowed with a metal vapor deposit to aid in interpretation of heights of surface irregularities.

A related instrument that has become a most important analytical tool in the ceramic industry is the scanning electron microscope (SEM). This instrument uses X rays or electrons scattered back from the surface "illuminated" by a rastered electron beam to generate an image with remarkable three-dimensional qualities. Since the beam need not be transmitted, the specimen does not have to be extremely thin, and replication procedures are not necessary. Figure 12.4 shows an ordinary transmission electron micrograph of a replica and a photographic image produced by a scanning electron microscope. The maximum magnification for a standard SEM is about $75,000 \times$. Some of the newer research instruments routinely produce $200,000 \times$.

Scanning electron microscopes have a number of analytical features which help interpret the superior visual images obtained of the specimen. For example, X rays generated as the electrons impact the sample provide a characteristic fluorescence pattern related to the elements present. (See X-ray analysis in next section.) The interpretation of these X rays yields a semiquantitative chemical analysis of the specimen at the point where the beam is focused.

Nonmicroscopic analysis

X-ray analysis

X rays make up the part of the electromagnetic spectrum having wavelengths shorter than ultraviolet radiation but longer than gamma rays. These very energetic waves are invisible to the eye, but they can be detected with photographic film or with special electronic detectors. Several important analytical methods that involve the interaction of X rays with material are in common use in the field of ceramics. While a detailed discussion of these methods is too involved to be included in this book, the frequency of use warrants a brief description of these techniques here.

When a beam of X rays strikes a crystalline solid, an interference pattern is produced. This pattern is generated by diffraction of the X rays by the planes of atoms in the crystal in much the same way that a diffraction grating produces an interference pattern with a beam of light. Because the exact interference (diffraction) pattern is unique for every crystalline substance, X-ray diffraction has become a standard method for identifying crystalline materials.

Each crystalline material contains many different families of parallel atomic planes. The "d spacing" or interplanar separation distance

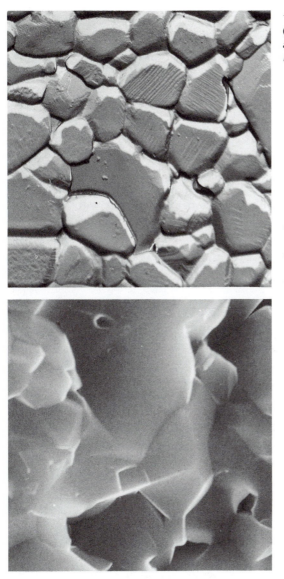

12.4. Electron micrographs (*courtesy V.E. Wolkadoff and R.E. Weaver, Coors Porcelain Co.*):

(a) Transmission electron micrograph of a replica of a thermally etched surface of alumina (10,000×).

(b) Scanning electron micrograph of a fracture surface of alumina (10,000×).

between individual planes in a particular family is uniform and is characteristic of the structure of the material. Figure 12.5 helps explain the presence of different families of atom planes (*a* and *b*), each having its own interplanar spacing (d_a and d_b).

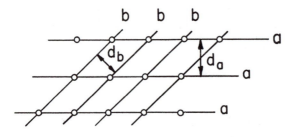

12.5. Schematic
representation of atomic
planes and interplanar
spacings in a crystal.

If a crystal is bathed in a beam of X rays and is slowly turned so that one after another of the families of atomic planes rotates through the beam, the X rays will be absorbed most of the time, but occasionally a beam will be diffracted back out from the crystal. The special conditions that must be met for this diffraction to take place are given by the Bragg equation:

$$d = \frac{\lambda}{2 \sin \theta} \tag{12.1}$$

Here λ is the wavelength of the X rays and θ is the angle between the entering beam (or the diffracted beam) and the diffracting crystal planes. The value of d calculated from the Bragg equation is the interplanar spacing for the particular set of planes responsible for diffracting the beam.

The X rays used in diffraction work are usually all of one wavelength, thus making Equation 12.1 easy to apply. The specimen is generally in the form of a fine powder in a rotating holder bathed in the beam. The angles at which diffraction occurs are determined either with a moving detector or with a piece of photographic film encircling the specimen.

Since each crystalline material has a unique set of d spacings, which can readily be found in tabulations such as the Powder Diffraction File (PDF) issued by the Joint Committee of Powder Diffraction Standards, the details of the diffraction pattern allows identification of the material. If the specimen is a mixture of minerals, the job of identification is more complicated but can be done by an experienced worker using the PDF and a computer.

A second analytical method employing X rays permits the determination of chemical composition of a material regardless of whether it is crystalline or glassy. It is especially applicable for rapid, routine analysis of material that does not vary greatly from day to day. When an intense beam of X rays strikes a material, it often excites the material itself to emit X rays. This X-ray fluorescence is not the same thing as diffraction, because the emitted X rays come out at all angles to the entering beam. Most important, however, is the fact that the wavelengths of the emitted X rays are characteristic of the kind of chemical elements in the specimen, and the intensity of emission at each characteristic wavelength is a measure of the amount of each element present. A rapid chemical analysis of a small specimen can be performed by simply analyzing the wavelengths and the intensities of the fluorescent X rays emitted.

Working with X rays and their generating equipment should be done only by carefully trained personnel fully aware of the hazards involved. X rays cause serious burns, and the dangers of permanent tissue damage by radiation overdose are very real. These dangers are compounded by the fact that X rays cannot be seen or felt, and thus a worker may be exposed to an intense beam and not realize it. X-ray equipment is always fitted with radiation shielding which must be kept in place during operation. It is absolutely essential that a portable radiation detector be kept on hand to survey the area during operation to detect X-ray "leaks" from the equipment. A second hazard in the operation of X-ray equipment is the high voltage on the X-ray generator tube and the detector.

Electron microprobe analysis

The electron microprobe analyzer provides a method for chemical analysis of very small areas within a larger solid specimen. This instrument uses an intense beam of electrons to excite fluorescent X-ray emission from the specimen. This beam can be focused down to a diameter of a few micrometers and can be moved across a polished surface of the specimen to analyze specific spots or features within the microstructure. As with X-ray fluorescence, the emitted X rays are analyzed for wavelength (corresponding to kind of chemical element) and intensity (corresponding to concentration of each element). Microprobe analysis is not commonly available in ceramic industrial plants, but most consulting laboratories and universities have access to the necessary instruments and will generally perform an analysis on a consulting basis. To some extent, the capability to perform elemental analyses in a scanning electron microscope has replaced the need to use these rather expensive instruments.

Spectrographic analysis

Conventional wet chemical analysis is still an important procedure in the ceramic industries (see, for example, ASTM C 573–81 [1990]), but when rapid determination of low concentrations of elements is required, a spectrographic analysis is often performed. A small pulverized specimen, highly excited in an electric arc, emits a spectrum of visible light characteristic of its chemical composition, and this spectrum can be recorded on photographic plates. Analysis of the spectrum is performed by wavelength (element) and intensity (concentration) measurements.

Differential thermal analysis

Clays and other raw materials that change composition or structure on heating may be identified by comparing the temperature differences that develop when the specimen and a nonchanging standard material (often aluminum oxide) are slowly heated side by side at the same controlled rate. The specimen absorbs heat (endothermic reaction) or expels heat (exothermic reaction) when it undergoes structural changes, decomposes, or melts. Since the standard material undergoes no such changes on heating, an endothermic reaction will cause the specimen to remain cooler than the standard material until the reaction is completed, while an exothermic reaction will cause the specimen to temporarily heat more rapidly than the standard material. Typically the temperature difference between the specimen and the standard material is determined by thermocouples and is plotted versus temperature of the standard material. The reactions appear as peaks or "inverse" peaks — the endothermic peaks being opposite in direction to the exothermic peaks. This procedure is called differential thermal analysis (DTA), and the plot is called a thermogram.

Many specific clays and other materials have been subjected to DTA, and unknown materials can sometimes be identified by comparison of their thermograms with this collection of data. Figure 12.6 shows a modern thermal analyzer capable of DTA and other types of thermal analysis. DTA data is frequently obtained simultaneously and correlated with gravimetric (weight) changes occurring in the specimen during the heating cycle. This time/temperature weight-loss procedure is called thermogravimetric analysis (TGA).

12.6. DTA and TGA apparatus (*courtesy Harrop Inc.*):

(a) Simultaneous DTA/TGA analyzer.

(b) Control and digital data acquisition system for a combined DTA/TGA system.

13 Strength

THE MEASUREMENT OF STRENGTH is a basic procedure in most ceramic industries and is often a routine quality control procedure in ceramic plants. The strength of a particular ceramic may indicate whether it has been properly fired. For example, in alumina ceramics, overfiring causes excessive growth of the alumina grains, which results in weakening of the ceramic. Underfiring results in excessive porosity and weak bonding in the ceramic and a subsequent reduction in strength.

The strength of a material is the level of stress at which it fractures. Since materials break at different stress levels under pressing loads (compressive strength) than under either stretching loads (tensile strength) or bending loads (modulus of rupture or flexural strength), it is necessary to specify the testing procedure used. Important strength-related properties of ceramics include the relationship between stress level and recoverable deformation (modulus of elasticity) and the rate of permanent deformation at high temperature (creep rate). Some methods for measuring these properties will be discussed in the following sections.

Modulus of rupture

The modulus of rupture (MOR) is the fracture strength of a material under a bending or fluxural load. The determination of the MOR is the simplest strength measurement made on ceramic materials, and for

measurements

this reason, it is a common quality control test.

The MOR measurement is made on a long bar of either rectangular or circular cross section, supported near its ends, with a load applied to the central portion of the supported span. There are two methods for applying the load: three-point loading and four-point loading. Figure 13.1 shows these two different configurations. In order to yield correct results, a bar in three-point loading must fracture at the exact center, but a bar in four-point loading can fracture at any point between the inner two loading points and still provide a valid measurement. Because of this less-restrictive fracture requirement, four-point loading is preferred. Fracture always originates at the surface carrying the tensile stress (the bottom surface when loaded as in Fig. 13.1).

For three-point loading using specimens of rectangular cross-section, the MOR is given by

$$\text{MOR} = \frac{3PL}{2bd^2} \tag{13.1}$$

where P is the load required to break the specimen, L is the span (distance between the outer supports), b is the width of the specimen, and d is the depth of the specimen. The usual units are psi (lb/in.^{2}), ksi (1000 lb/in.^{2}), or MPa.

For three-point loading using cylindrical specimens, the MOR is given by

$$\text{MOR} = \frac{8PL}{\pi D^3} \tag{13.2}$$

where D is the diameter of the specimen.

229

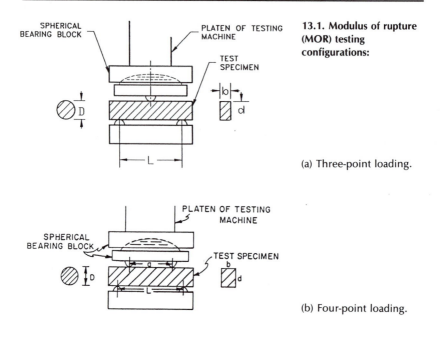

13.1. Modulus of rupture (MOR) testing configurations:

(a) Three-point loading.

(b) Four-point loading.

For four-point loading using rectangular specimens, the MOR is given by

$$\text{MOR} = \frac{3P(L - a)}{2bd^2} \tag{13.3}$$

where a is the distance between the inner two load application points.

For four-point loading using cylindrical specimens, the MOR is given by

$$\text{MOR} = \frac{8P(L - a)}{\pi D^3} \tag{13.4}$$

The most important parameters to control in MOR determinations are the rate of loading (especially in tests made above room temperature), the ratio of span-to-specimen thickness (L/d), and the specimen alignment. The ratio of span-to-specimen thickness must be at least 10 to 1 or corrections to Equations 13.1–13.4 must be made. The specimen cannot twist while being loaded. Standard testing methods should be used for all strength measurements. A partial list of ASTM methods of testing for MOR of ceramics is given in Table 13.1.

TABLE 13.1. ASTM standard methods of test for MOR of ceramics

Method	Applicable Ceramics
C 93–84	Insulating firebrick
C 133–84	Cold refractory brick
C 491–85	Air-setting plastic refractories
C 583–80 (1990)	Refractories at high temperatures
C 158–84 (1989)	Glass
C 674–88	Whitewares
C 67–90a	Building brick

Figure 13.2 shows a modern testing machine that can be used for measuring the strength of green or fired ceramics. Testing machines should meet ASTM specifications E 4–89 and E 74–91.

13.2. Machine used to determine strength. This model is microprocessor controlled and provides digital readouts of both load and strain. Tensile grips are shown in place, but other configurations allow testing in compression and flexure (*courtesy Instron Corp.*).

Compressive strength

The compressive strength of a material is a measure of its ability to bear crushing or pressing loads. The compressive strengths of ceramics are much higher than their tensile strengths. Since brittle materials such as ceramics normally break in tension, a true compressive strength is usually very difficult to measure.

The general configuration for this test is shown in Figure 13.3. The compressive strength is given by the equation

$$S_c = \frac{P}{A} \tag{13.5}$$

where P is the load at fracture and A is the cross-sectional area of the test specimen.

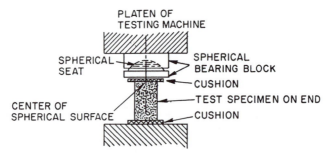

13.3. Compressive strength test configuration.

The measurement of compressive strength requires more care in specimen preparation and alignment than the measurement of MOR. In particular, the specimen faces bearing the load must be absolutely flat and parallel. If these criteria are not met, the load will be carried unevenly by the specimen, causing failure at low loads and thus giving apparently low compressive strengths. Special caps or cushioning materials are often used to distribute the load uniformly over the bearing surfaces. These cushions might consist of thin mild steel wafers against a hard, sintered specimen.

The specimen size, configuration, and load rate all influence the measured compressive strength. A partial listing of ASTM methods of test for compressive strength of ceramics is given in Table 13.2.

TABLE 13.2. **ASTM standard methods of test for compressive strength of ceramics**

Method	Applicable Ceramics
93–84	Insulating firebrick
C 133–84	Refractory brick
C 773–88	Fired whitewares
C 579–82	Chemical-resistant mortars
C 67–90a	Brick and structural clay tile

Tensile strength

The tensile strength of a material is a measure of its resistance to fracture under a stretching or pulling load. Elaborate specimen preparations and specimen alignment procedures are required to obtain satisfactory results with ceramics or other brittle materials. The specimen gripping devices must be carefully designed to avoid creating undesirable shear or bending stresses in the specimen.

Figure 13.4 illustrates one method for obtaining tensile strengths; more elaborate gripping arrangements are often used. The tensile strength is given by

$$S_t = \frac{P}{A} \tag{13.6}$$

where P is the load at fracture and A is the cross-sectional area of the specimen at the point of breaking. ASTM Standard Method of Test C 565–83 is often used when testing tensile strength of ceramics.

13.4. Tensile strength testing:

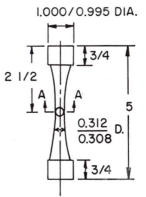

(b) A tensile test specimen.

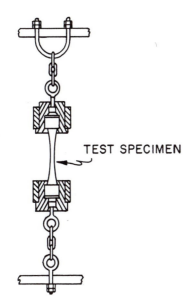

(a) A tensile test configuration.

Modulus of elasticity

The modulus of elasticity (Young's modulus) of a material is the amount of stress needed to produce a unit of strain and is thus a measure of its rigidity. This modulus will usually be very high for most ceramic materials. This property can be determined in several ways, but for ceramics it is determined almost exclusively by sonic methods. These procedures are based on the principle that the modulus of elasticity of a specimen is proportional to the square of a particular natural vibrational frequency (or resonant frequency) of the specimen. The resonant frequency of ceramic materials can be measured electronically (see ASTM C 623–71 [1989]) and Young's modulus can thus be determined. In quality control operations, the modulus of elasticity of a part may not be calculated, but instead, standards are set up for resonant frequency, and parts that have frequencies below a set limit are either refired or rejected.

It has been a common practice for centuries to tap ceramic ware to see if a "good ring" results; this effect is related to modulus of elasticity. A dull thud in a particular ceramic part (which usually has a ring) indicates that it does not possess its expected high modulus, probably indicative that there is a crack in the part or that it has not been fired to maturity.

Sonic testing is also a valuable tool for determining the relative thermal shock resistance of porous ceramics (see Chapter 14).

Hardness

Great hardness is one of the most characteristic properties of ceramics, but because of the brittleness of ceramic materials, hardness is also one of the most difficult properties to measure. Several methods have been developed which give fairly reproducible results. One is based on the resistance of the ceramic to indentation. Usually, a diamond stylus is forced into the surface of a ceramic specimen under a standard load, and the depth of penetration is measured. The Alumina Ceramic Manufacturing Association has set up standards for hardness using the Rockwell 45N hardness scale—now designated ASTM D 785–89. The test is made on polished specimens employing a 45-kg load on the diamond stylus. Although the numerical difference between alumina samples of various compositions is small, the test results are quite reproducible. Figure 13.5 shows instruments for hardness testing.

13.5. Hardness testing machines:

(a) Hardness testing machine (*courtesy Rams Rockford, Inc.*).

(b) Image analysis–based, fully automated microhardness tester with programmable stage patterns, automated measurements and print-out of results and statistics (*courtesy LECO Corporation*).

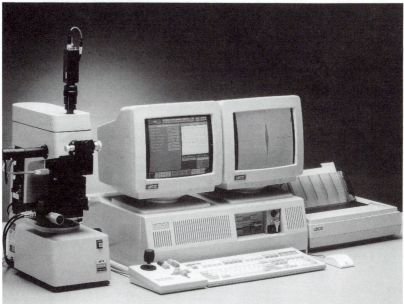

A second hardness testing method measures the resistance of a material to abrasion. The specimen is abraded for a set time by impinging abrasive grain onto its surface using high-pressure air. The hardness of the ceramic is determined by comparing with materials of known hardness. ASTM Standard Method of Test C 448–88 describes one such test for measuring the abrasion resistance of porcelain enamels.

Ceramic technologists sometimes refer to the Mohs scale of hardness used by mineralogists. This scale uses ten standard minerals, each of which will scratch all minerals below it on the scale. Ceramics are given a numerical rating on this scale by scratch trials with the mineral standards: (1) talc, (2) gypsum, (3) calcite, (4) fluorite, (5) apatite, (6) orthoclase (feldspar), (7) quartz, (8) topaz, (9) corundum, and (10) diamond.

Impact testing

Ceramics are often used in applications in which impact resistance is an important criterion. The impact resistance of a ceramic can be determined by measuring the minimum instantaneous fracturing load of a small ceramic bar, ball, or plate. This load is usually applied by a swinging pendulum and is measured in foot-pounds (see Fig. 13.6). Impact resistance is used as a quality control measure in the whitewares industry. (See ASTM Standard Method of Test C 368–88).

Creep test

Since many ceramics are required to carry loads at high temperatures, the resistance to permanent continuous deformation under such loading conditions (creep resistance) is an important property. Comparative tests between different materials are routinely run in the refractories industry. A creep test (hot load test) may be run using flexural, tensile, or compressive loading methods. In most cases the deformation under constant load at a given temperature is measured as a function of time. Deformation can be measured optically, but a linear variable differential transformer (LVDT) is generally more satisfactory.

After a rapid initial deformation, the creep rate usually gradually diminishes to a constant rate, which continues for very long periods of time; this is the rate of general interest in the refractories industry. The constant creep rate is usually reported as percent deformation per hour for tensile and compressive tests. The time to reach the constant creep rate is very short at relatively high temperatures and increases greatly as the temperature decreases.

13.6. Impact tester (*courtesy Testing Machines, Inc.*).

14 Thermal

THE DESIGN ENGINEER who uses ceramics in high-temperature applications needs to know how rapidly the ceramic will transfer heat, the quantity of heat it will absorb as the temperature increases, and the magnitude of the reversible size changes the ceramic will undergo on heating or cooling. The three properties that allow computation of these quantities are the thermal conductivity, the specific heat, and the coefficient of thermal expansion of the ceramic. These properties, along with thermal shock behavior and spalling resistance, are generally grouped together into a single category called the thermal properties of the material. The permanent changes occurring when a ceramic is first fired—decomposition, firing shrinkage, or densification, for example—are discussed in some detail in Chapter 5.

Measurements of thermal properties are usually difficult to perform and require careful control and special techniques. The significance of the properties is discussed in the following sections along with the principle of the measurement. Details of specific procedures can be found in standard test methods.

Specific heat

The amount of heat energy that must be absorbed by a unit weight of a material to raise its temperature one degree is called the specific heat of the material. The specific heat of water is used to fix the size of two

property measurements

standard heat quantities; the British Thermal Unit (BTU) is the amount of heat required to raise the temperature of 1 pound of water by 1°F, and the calorie is the amount of heat required to raise the temperature of 1 g of water by 1°C. Conventional units on specific heat are either BTU/lb − °F or cal/g − °C.

The heat capacity of a specimen is the total amount of heat energy necessary to raise the temperature of the specimen one degree. The heat capacity is not a property of the material; the relationship between heat capacity and specific heat is given in Equation 14.1.

$$\text{Heat capacity} = (\text{specific heat})(\text{weight of specimen}) \qquad (14.1)$$

Very often, tabulations are made of molar heat capacity of materials — that is, the heat capacity for one mole of material. The relationship between specific heat and molar heat capacity is indicated by

$$\text{Specific heat} = \frac{\text{molar heat capacity}}{\text{molecular weight}} \qquad (14.2)$$

The measurement of specific heat is usually accomplished with a device called a calorimeter, which uses a known amount of water to absorb heat from a weighed hot specimen dropped into it. When the temperature change of the water is measured with a sensitive pyrometer, the known specific heat of water allows calculation of the amount of heat absorbed. When this procedure is repeated using specimens at a series of temperatures, the specific heat of the specimen material as a function of temperature can be calculated readily.

An instrument frequently used in the analysis of raw materials,

239

polymers, and glazes is the differential scanning calorimeter (DSC), shown in Figure 14.1. The traditional instrument heats a sample and a reference material in the same furnace using thermocouples to measure the temperature difference between the two. The energy changes in the sample are calculated using heat flux equations. A more accurate instrument uses two furnaces, one for the sample and one for the standard material. Minute temperature changes can be measured accurately with platinum resistance thermometers. The temperature in both furnaces is kept the same by supplying or removing energy from one or both furnaces as changes occur in the sample. The amount of power required to maintain temperature equality is directly proportional to the heat absorbed and released during changes in the sample. Heat flux equations are not required. The instrument must be calibrated using calibration standards over the full temperature range of interest.

DSC data are often compared with DTA and TGA data (see Chapter 12) for the same material.

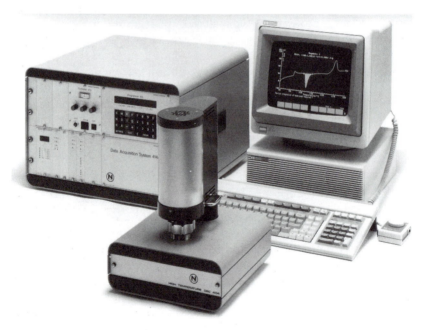

14.1. Computer-controlled, differential scanning calorimeter (DSC) (*courtesy Netzsch, Inc.*).

Thermal conductivity

Heat transfer within a material can be accomplished by three different mechanisms: (1) the flow of heat through a solid ceramic material usually occurs by conduction; (2) if the specimen is porous, a small part of the heat may be transferred by moving gases within the porosity through convection; and (3) at higher temperatures, heat may be transferred across pores by radiation.

Thermal conductivity is a property that describes the rate of heat transfer through a material by all mechanisms combined. Thermal conductivity is measured by determining the rate of heat flow through and the temperature drop across a carefully insulated piece of material heated from only one side. The thermal conductivity is calculated by the equation

$$k = \frac{Qx}{A(T_2 - T_1)} \tag{14.3}$$

where Q is the rate of heat flow through a specimen of cross-sectional area A and thickness x when there is a temperature difference $(T_2 - T_1)$ across the thickness. Temperature T_1 is often kept at a fixed "reference" temperature.

Since the thermal conductivity may vary considerably with temperature, Equation 14.3 may have to be used over a number of different $(T_2 - T_1)$ temperature intervals so that an accurate plot of k versus average temperature can be made.

Ceramics of both low and high thermal conductivities have important uses. Those of low conductivity, known as thermal insulators, make excellent heat barriers in furnace walls and crowns. Those of high thermal conductivity, such as dense BeO, SiC, Si_3N_4, AlN, and graphite, help remove heat from critical areas in electronic and other devices.

The thermal conductivity of a ceramic, particularly that of a good thermal insulator, is difficult to measure. Several methods have been accepted by ASTM and are listed in Table 14.1.

TABLE 14.1. **ASTM standard methods of test for thermal conductivity of ceramics**

Method	Applicable Ceramics
C 201–86	Refractories (includes description of apparatus)
C 417–86	Castable and plastic refractories
C 202–86	Firebrick
C 182–88	Insulating firebrick
C 408–88	Whitewares

Thermal expansion

Like all materials, ceramics typically reversibly expand when heated and contract when cooled. If allowance is not made in a part or structural design for these size changes, high stress levels and probable failure will result during temperature changes. The magnitude of the dimensional change on heating is directly proportional to the original length, the change in temperature, and a property of the material called the mean coefficient of linear thermal expansion. The thermal expansion coefficient is a mean measure of length change taking place between room temperature and some elevated temperature. The volume thermal expansion coefficient of a material is approximately three times its linear thermal expansion coefficient, provided that the material is isotropic (exhibits uniform expansion in all directions).

The determination of linear thermal expansion coefficients involves the measurement of small length changes as a function of temperature. Several methods are available for making these measurements, including interferometry (ASTM C 539–84 [1990]) and the use of dilatometers (ASTM C 372–88). The dilatometer may employ either a dial gauge or a linear variable differential transformer (LVDT) to measure total expansion (Fig. 14.2).

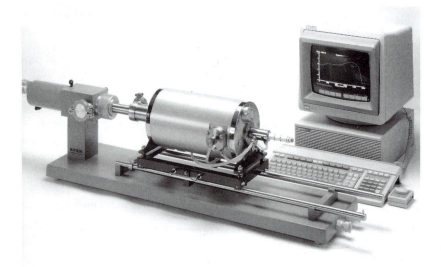

14.2. Research dilatometer for thermal expansion coefficient measurements (*courtesy Netzsch, Inc.*).

A traveling telescope mounted on a micrometer movement and sighted through furnace windows can also be used to observe the change in length of a heated specimen, which is supported horizontally in the furnace in a nonrestrained manner. A bed of free-flowing refractory grain is a suitable support for this procedure.

Thermal shock and spalling resistance

When a ceramic specimen is subjected to extremely rapid heating or cooling, large temporary differences in temperature may develop between the surface of the specimen and its interior, causing stresses which often result in fracture. This condition, known as *thermal shock,* will usually result in general failure of small parts. In more massive parts, the fracture may cause pieces of the surface to break away; this phenomenon is called *spalling.*

It is not generally possible to assign a numerical value to the thermal shock resistance of a material, but only to compare different materials in their resistance to shock. As a general rule of thumb, the thermal shock resistance of a material improves with an increase in strength and thermal conductivity and with a decrease in thermal expansion coefficient and modulus of elasticity. Microstructure, especially the presence of crack-arresting porosity, also has a strong influence on the resistance to failure of a part by thermal shock.

Methods for measuring thermal shock resistance (Table 14.2) generally employ a reproducible scheme for shocking a series of specimens, with a measurement of some sensitive property being made after each subsequent shock cycle. Shocking is usually accomplished by quenching a heated sample, either with an air blast or by dropping it into water or oil. Either the modulus of elasticity (sonic) or modulus of rupture of the specimen can be measured after shocking, with the decrease in these properties due to increased cracking after each thermal-shock cycle being used as the measure of shock resistance. A second measure of thermal shock resistance sometimes used is the number of shock cycles necessary before visible failure of the part.

TABLE 14.2. **ASTM standard methods of test for thermal shock resistance of ceramics**

Method	Applicable Ceramics
C 554–88	Glazed whitewares
C 385–58 (1983)	Porcelain enameled utensils
C 484–66 (1981)	Glazed ceramic tile
C 600–85	Glass pipe
C 149–86	Glass containers

The resistance to spalling away of material from a rapidly quenched refractory surface is sometimes measured by a panel test (see Table 14.3). In these tests, refractory bricks are used to construct a panel lining for a furnace door. After the furnace has been heated to some elevated temperature, the door is swung open, exposing the refractory panel to regulated cold air and water blasts. The spalling damage is noted as weight loss after a series of consecutive heating and quenching cycles.

TABLE 14.3. **ASTM standard methods of test for spalling resistance of refractories**

Method	Applicable Refractories
C 38–89	Refractory brick
C 107–89	High duty fireclay brick
C 122–89	Super duty fireclay brick

15 Temperature

MANY INDUSTRIES are concerned with the measurement and control of temperature, and this is especially true of the ceramic industries, where high-temperature treatments play such an important role in the overall processing of every product. The ability to control temperature implies the ability to measure it accurately. The branch of technology called pyrometry is concerned with the measurement of temperature. This chapter describes the most common pyrometers used to measure temperatures in the ceramic industries; further information can be found in pyrometry texts such as McGee (1988). The complex topic of temperature control is beyond the scope of this book.

Temperature scales

The temperature of an object or an environment can be expressed on any one of several familiar temperatures scales. The Fahrenheit scale has long been the most popular temperature scale in the United States. On this scale, the standard boiling point of water is 212°F, and the freezing point is 32°F. The other common temperature scale widely used by industry and by scientists around the world is the Celsius (formerly centigrade) scale, in which the standard boiling point of water is 100°C and the freezing point is 0°C. It is often necessary to convert temperatures from one scale to the other; Table A.2 in the Appendix provides for this conversion. The two scales are identical at a temperature of −40°; therefore, a simple three-step scheme for converting between scales is:

1. Add 40° to the temperature reading to be converted.

measurements

2. Multiply this sum by 5/9 if converting from Fahrenheit to Celsius or by 9/5 if converting from Celsius to Fahrenheit.
3. Subtract 40° to yield the converted reading.

Two widely used absolute temperature scales have their zero points at the absolute zero of temperature, but have units of different sizes. The unit on the Kelvin absolute scale is the same size as the degree on the Celsius scale; however, rather than calling that unit a degree, it is simply called a kelvin (K). Water boils on the Kelvin scale at 373 K and freezes at 273 K. (These temperatures are rounded to the nearest unit, as are the temperatures cited on the Rankine scale.) The Rankine absolute scale has the same size degree as the Fahrenheit scale. Water boils on the Rankine scale at 672° R and freezes at 492° R. The conversions from common scales to absolute scales are

$$K = °C + 273$$

$$°R = °F + 460°$$

Negative temperatures are not unusual on the common scales, but only positive temperatures can occur on the absolute scales. Table 15.1 gives a comparison of temperatures on the four scales for familiar situations.

TABLE 15.1. **Comparisons of familiar temperatures**

	° F	° C	° R	K
Dry ice	−110	−79	350	194
Freezing point of mercury	−38	−39	422	234
Room temperature	75	24	535	297
Melting point of aluminum	1220	660	1680	933
Firing porcelain	2102	1150	2562	1423
Melting point of glass	2642	1450	3102	1723
Melting point of pure iron	2795	1535	3255	1808

Thermocouples

By far the most common pyrometer in use in the ceramic industries is the thermocouple, which consists of two uniform wires of different composition welded together at one end to form a junction and attached to a voltage-measuring instrument. If the temperature of the welded junction is different from the temperature where the wires connect to the measuring instrument, a small reproducible dc voltage is generated in the circuit. The magnitude of the voltage generated is proportional to the difference in temperature of the two ends of the circuit—that is, the greater the temperature difference between the ends, the higher the voltage. Once it has been properly calibrated, the thermocouple can provide an accurate measure of the difference in temperature of the two ends of the circuit.

The welded end of the circuit is called the measuring junction, and the connections at the voltage measuring instrument constitute a second junction called the reference junction. Figure 15.1 shows a schematic of the circuit of the thermocouple just described. The voltage measured will be proportional to the difference in temperature between the measuring junction and the reference junction. If the temperature of the reference junction is fixed and known, the voltage generated is a direct measure of the temperature of the measuring junction.

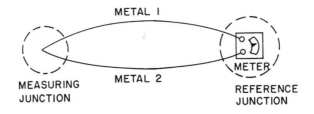

15.1. Basic thermocouple circuit.

Often the thermocouple is constructed with a second welded junction as in Figure 15.2. When the two metals are connected as shown, the second welded junction becomes the reference junction, and the thermocouple behavior is independent of the presence of the connections at the measuring instrument. The voltage read at the instrument is now proportional to the difference between the temperatures of the two welded junctions, and the temperature at the instrument is of no consequence to the circuit. In this configuration, the temperature of the reference junction can be fixed, for example by immersion in an ice bath, at 32°F or

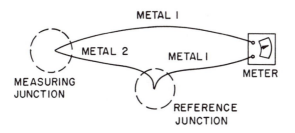

METAL I

METAL 2   METAL I

METER

MEASURING
JUNCTION

REFERENCE
JUNCTION

15.2. Thermocouple circuit with welded reference junction.

0°C, and variations in room temperature at the measuring instrument will not influence the behavior of the thermocouple.

Any two different metals could be used to construct a thermocouple, but of the thousands of possibilities, only three combinations are commonly used in the ceramic industries, with perhaps an additional three to five combinations in limited use. The iron-constantan or type J thermocouple is made from one pure iron wire and one wire of a copper-nickel alloy called constantan. This couple has a useful temperature range from about −300° to +1400°F (−184° to +760°C) with good oxidation resistance in air up to about 750°F (400°C). It is useful in dry, sulfur-free reducing atmospheres over its entire temperature range. A related couple, the copper-constantan or type T couple, has a narrower useful range in air, −300° to +650°F (−184° to +343°C), but because of its improved resistance to moisture attack, it is often used in favor of the type J couple in high-humidity driers.

The chromel-alumel or type K couple is made from two alloys high in nickel. It has a useful temperature range from −300° to +2200°F (−184° to +1204°C) with a short-time maximum-use temperature of 2450°F (1343°C). This couple must *always* be used in oxidizing conditions or it will rapidly fail. Sulfurous gases also cause rapid deterioration. Because of its fairly high-temperature capability in air, the type K couple finds wide usage in ceramic kilns for firing enamels and ware that mature at low temperature.

The most important thermocouple in use by the ceramic industries is the type S, a couple using pure platinum as one leg of the circuit and an alloy of 90% platinum and 10% rhodium as the other leg; this couple is also known as the Pt-Pt 10% Rh couple. The useful range for this couple is 32° to 2650°F (0° to 1454°C), with short-time usage at 3100°F (1704°C) possible. The type S couple must *always* be operated in sulfur-free oxidizing atmospheres, or rapid deterioration will occur. To be on the safe side, this expensive couple is normally encased in a gas-tight protection tube whenever it is to be used in a combustion furnace. The voltage generated by this couple is much lower than for the type K couple at the same temperature, but its wider temperature capability makes the type S couple popular for most ceramic kiln applications. A related couple, the type R, having a pure platinum leg and an 87% platinum plus 13% rhodium leg, is sometimes encountered; since its temperature-voltage characteristics are different from the type S, the two should never be used interchangeably.

Because the voltage generated by a thermocouple with its junctions at fixed temperatures depends only on the compositions of the two metals used for constructing the couple, it is possible to prepare tables of voltages for each type of couple that will be valid for every couple of that type. Tables A.3 through A.7 in the Appendix list characteristic output voltages for types J, T, K, S, and R couples. These tables are valid only when the reference junction of the couple is held at 32°F (0°C) as in an ice bath. For example, referring to Table A.6, it is seen that every type S thermocouple with its reference junction at 32°F (0°C) and its measuring junction at 1500°F (816°C) will generate 7.498 millivolts (mv), regardless of the length or diameter of the wires, or any other construction feature. Temperatures are determined by measuring the output voltage of the thermocouple and searching the proper table until that voltage is found. Linear interpolation is generally acceptable between adjacent readings in the tables for all but the most accurate measurements.

These tables can also be used to convert observed thermocouple voltage to measuring junction temperature even if the reference junction temperature is not 32°F (0°C), provided that a simple reference junction correction is first made. The method for making these corrections consists of the following steps:

1. Measure the output of the couple to obtain the uncorrected voltage.
2. Measure the actual temperature at the reference junction with a thermometer or by other means.
3. In the correct thermocouple table, read the voltage corresponding to the reference temperature measured in step 2. This will be a positive voltage if the reference junction is above 32°F and negative if the reference junction temperature is below 32°F.
4. Add this voltage (step 3) to the uncorrected voltage (step 1) to obtain the corrected voltage of the couple.
5. Enter the table again and read the temperature corresponding to the corrected voltage; this is the actual measuring junction temperature.

This reference junction correction procedure must always be used when no ice bath is employed, and most especially when the reference junction is composed of the connections at the instrument (Fig. 15.1). Many instruments that automatically indicate and record temperature from a thermocouple also automatically make these reference junction corrections; other instruments have manually adjustable internal voltages to accomplish these corrections.

Millivoltmeters and potentiometers

Two different types of instruments are used to measure the small dc voltages generated by thermocouples—the millivoltmeter and the potentiometer. Each has certain advantages in its use, but each also requires certain precautions.

A millivoltmeter indicator is simply a dc voltmeter that indicates the difference in voltage across its terminals. Its main convenience is that it indicates visually and directly the voltage being measured. If only one kind of thermocouple is to be used with the meter, the dial can be calibrated to read directly in temperature rather than in millivoltage. Analog meters have been in use for many years; today most millivoltmeters provide a digital readout (Fig. 15.3).

The chief drawback of a millivoltmeter for use with a thermocouple arises from the fact that current flows through the thermocouple during the course of the measurement. Since the thermocouple circuit possesses some electrical resistance, the current flow will result in a smaller voltage drop across the meter than that actually generated by the couple, and an erroneously low temperature reading will result. Compensation for this resistance effect can be accomplished by insuring that the couple in use with the meter has a specific value of resistance, and calibrating the meter accordingly. A meter calibrated in this way carries an "external

15.3. Digital **millivoltmeter** for thermocouple temperature measurement (*courtesy Leeds & Northrup Co.*).

resistance" specification on its nameplate, and only couples having this resistance should be used with that particular meter.

The potentiometer uses a small, internally generated dc voltage to oppose the voltage generated by the thermocouple. When the two opposing voltages are equal, no current will flow in the combined circuitry of the couple and the instrument. Since no current is flowing in the circuitry when the potentiometer has achieved balance, the external resistance of the thermocouple can be disregarded, and a true reading of the couple voltage is always obtained. Because of the complexity of such instruments, potentiometers are more expensive than millivoltmeters.

Radiation pyrometers

All objects radiate a spectrum of electromagnetic energy to their surroundings. As the temperature of an object rises, the average wavelength of radiated energy decreases and the intensity of radiated energy increases. An object at 800°F (427°C) radiates energy at wavelengths too long to be seen by the eye, but this infrared energy can be "felt" by the hand up to several feet away. If the object is further heated to about 1200°F (649°C), the average radiated wavelength will become shorter, and some energy will be radiated as visible light at the red end of the visible spectrum; the object is now said to be incandescent. As the temperature is further raised, the average wavelength shortens more and more, so that the visible part of the radiated energy changes in hue from deep red, through orange to yellow, and finally reaches white at about 2200°F (1204°C). At the same time the intensity of radiation rises rapidly, so that by the time the object appears white, it is too brilliant to be observed without filters to protect the eye.

This characteristic radiating behavior of heated objects can be used to determine their temperatures if suitable instruments can be devised to follow the changes. With a little experience, the unaided eye can serve as a crude pyrometer by observing the color shifts accompanying heating. Accurate radiation pyrometers are based on optical detection of the change in intensity of the energy radiated as the object temperature changes. These instruments would not be very useful unless all objects at the same temperature radiated at the same intensity. Because this is not generally the case, a correction factor called the emittance must be applied to all measured intensities so that all objects will have the same relationship between corrected intensity of radiation and object temperature. When the proper emittance correction is made, a radiation pyrometer can measure the temperature of any object, no matter what its particular radiation pattern may be. The emittance of a particular object is a function of the material, the angle of view, the surface finish, and other parameters. The details of the calculations involved in making emittance corrections depend on the type of pyrometer being used and are beyond the scope of this book.

Many years ago it was discovered that all opaque objects (no matter what their particular emittance might be), when placed inside an enclosure and viewed through a small hole in the enclosure wall, behave as if they have an emittance of one (the maximum possible value). Objects with unit emittance are called *black bodies,* and they radiate the maximum possible amount of energy at all temperatures. The black body cavity arrangement for viewing objects eliminates the need to apply emittance corrections and is therefore used whenever possible. A practical realization of a black body cavity consists in observing objects inside a furnace through a small peephole in the furnace wall. To insure the achievement of black body conditions, the ratio of the distance of the object from the furnace wall to the peephole diameter should be at least 8 to 1.

The total radiation pyrometer (Fig. 15.4) is a common instrument used in the ceramic industries, especially the glass industry. It receives energy from a wide band of the spectrum radiated by a hot body and focuses it onto a sensor, usually either a thermopile (several thermocouples in series) or a photocell. Either sensor will generate a millivoltage that can be measured by a suitable instrument, and by the use of suitable tables, translated into object temperature. The pyrometer is usually set up to view objects through a hole in the furnace wall or crown so that black body conditions are achieved; otherwise, emittance corrections would be necessary. An important source of error in using total radiation pyrometers for objects inside furnaces arises from the selective absorption of radiant energy at certain wavelengths by combustion product gases (especially CO_2 and H_2O) in the furnace atmosphere. To avoid the erroneously low temperature reading that would result, total radiation pyrometers are often fitted with long, refractory, closed-end tubes that extend into the hot part of the furnace. The sensor receives energy only from the inside of the end of the tube, and since no combustion product gases can enter the closed tube, no absorption errors result. Radiant images can be transmitted efficiently by fiber optics from remote locations to control instrumentation for electronic analysis.

The spectral radiation pyrometer, more commonly called simply the optical pyrometer, measures the intensity or brightness of the energy radiated by a hot object at a single wavelength instead of across a wide spectrum band. The brightness determination is generally made by com-

**15.4. Total radiation
pyrometer** (*courtesy Leeds
& Northrup Co.*).

paring the brightness of the hot object with the brightness of a filament
in a special internal electric lamp. The pyrometer (Fig. 15.5) is constructed in such a way that the image of the object is formed inside the
pyrometer tube at the location of the lamp filament. When the operator
looks through the tube, the image of the filament is superimposed on the
image of the hot object. A red filter in the eyepiece of the pyrometer
assures that the operator sees only a narrow bandwidth of wavelengths
from both filament and object. A brightness match between filament
and object occurs when the two images blend together and the filament
seems to disappear against the background of the object; this kind of
behavior leads to these instruments sometimes being called disappearing
filament optical pyrometers. A brightness match is achieved by varying
the filament current, and a table of filament current versus temperature
is used to determine the object temperature. Very often, the filament
current rheostat carries a scale that permits direct reading of temperature. Various schemes for automated brightness matching have been developed; these add greatly to the cost of the pyrometer.

An experienced operator can achieve great precision with an optical
pyrometer unless the brightness of the object becomes too great. To
allow determination of very high temperatures, and to lengthen filament
life by limiting operating current, a series of range filters are provided
with each instrument. The optical pyrometer, like the total radiation
pyrometer, is subject to emittance errors unless black body conditions

15.5. Optical pyrometer.
While many are still
found in ceramic plants,
this instrument is no
longer made (*courtesy
Leeds & Northrup Co.*).

are achieved. Absorption errors, however, are not a serious problem
with optical pyrometers.

Because the instruments themselves never heat up, there is really no
upper temperature limit for either type of radiation pyrometer. A practical upper limit does exist, however, since both require a calibration table,
which can only be constructed using temperatures measured by some
other pyrometer. The object being viewed with an optical pyrometer
must be hot enough to be incandescent before the eye can make brightness comparisons. This results in a definite lower limit of usefulness for
the optical pyrometer. The total radiation pyrometer is less limited, since
any object above absolute zero radiates some energy. The lower practical
limit of the total radiation pyrometer occurs at about 200°F (93°C),
where the electrical signal generated by the sensor becomes so small that
accurate measurement is almost impossible.

Optical pyrometers, while still commonly encountered in ceramic
plants, are gradually being displaced by other pyrometer types.

Pyrometric cones

The use of *pyrometric cones* is unique to the ceramic industries.
Their use provides a measure of the combined effects of temperature and
exposure time during the firing operation. The cones themselves (Fig.

15.6) consist of a series of standardized unfired ceramic compositions molded into the shape of slender triangular pyramids. When the cones are heated, they undergo many of the processes that occur in a vitrifying ceramic — in particular, the progressive formation of liquid. When sufficient liquid has formed in the cone to cause it to slump under its own weight, it will bend over until the tip touches the mounting plaque. When this occurs, the cone is said to have reached its end point.

Various cones in the standard series are formulated to provide end points corresponding to reproducible amounts of *heat work* (the combined effect of temperature plus time). The end-point temperatures are

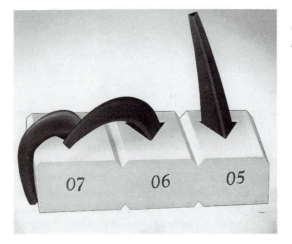

15.6. Pyrometric cones (*courtesy Edward Orton Jr. Ceramic Foundation*):

(a) Fired pyrometric cone plaque showing cone 06 at end point and guard cone 05.

(b) Orton self-supporting cones before and after firing, showing cone 06 nearing its end point.

quite dependent upon heating rate, so cones are not intended to provide an accurate measure of temperature. The most common series of cones are numbered consecutively in order of ascending end points from 022 to 01 and from 1 to 42, giving a total of 64 cone numbers. Several of these numbers are no longer made, and several have been added to the original series. The end point temperatures of standard pyrometric cones heated at standard rates are given in Table A.8 in the Appendix.

In use, several consecutively numbered cones are mounted in a supporting plaque at a carefully controlled angle of tilt. The cone whose end point represents the proper amount of heat work for the ceramic ware being fired is placed next-to-last in the plaque. The plaque is placed inside the kiln in a position that can be observed during firing. Identical plaques are often set at several locations throughout the kiln to test the uniformity of kiln behavior. During firing, the plaques are periodically observed through peepholes, and as the desired amount of heat work is neared, the lowest-end-point pyrometric cones deform and bend over to touch the plaque. When the desired cone reaches its end point (Fig. 15.6), the firing is complete. The last cone is called a guard cone and is used to indicate whether overfiring has occurred.

Thermocouples or other pyrometer instruments are normally used today to measure and control the heating schedule of a kiln, and pyrometric cones are widely used as a check on the firing of ceramic materials that mature by vitrification. A second use of pyrometric cones is to test the refractoriness of fire clays. In this application, a sample of the fire clay is molded into a smaller version of the standard cone shape, and it is heated along with a number of small standard cones so that its end point can be determined in terms of an equivalent cone number (Fig. 15.7).

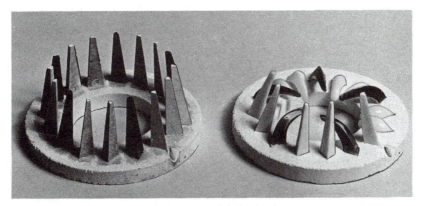

15.7. Pyrometric cone equivalent (PCE) test: left—unfired test plaque; right—fired plaque with test material (dark) cones at end point.

This number is then designated as the *pyrometric cone equivalent* (PCE) of that particular clay. (See ASTM method of test, C 24–89.)

Buller rings, which are unfired ceramic discs of a selected composition and geometry, are sometimes used in place of pyrometric cones in the whitewares industry to monitor kiln firings. The fired diameter of the disk is calibrated as a Buller ring number. Buller rings can be pulled from periodic kilns during the firing process to determine the status of the firing.

Miscellaneous pyrometers

Several additional pyrometers are employed occasionally in the ceramic industries. Included in this group are expansion pyrometers, which include the ordinary mercury-in-glass thermometer, as well as gas-filled and bimetallic strip pyrometers. Resistance pyrometers employ the change in electrical resistance of metals with temperature as the basis for measurement; and thermistors utilize the change in electrical resistance of semiconductors with temperature. Occasionally certain paints and resins which undergo definite irreversible color changes on heating to a specific temperature are employed, especially to detect a developing hot spot in the shell of a furnace, which could indicate impending refractory failure.

Appendix

Chart A.1. CONVERSION FACTORS

The traditional English system of units (inches, feet, pounds, etc.) is still in widespread use in the United States, but it is gradually being replaced by the SI (Systeme International) or metric system of units. One of the great criticisms of the English system is that different units for the same basic quantity (say length, where inches, feet, yards, and miles are all in common use) are related to each other in arbitrary ways. This forces one to remember a whole set of arbitrary conversion factors in order to manipulate quantities in the English system.

The SI system has long been the standard system used by the worldwide scientific community and by most industrial communities outside the United States. The SI system is a decimal system; that is, all units for a particular type of quantity (say, length) are derived from the basic unit (say, the meter) by multiplying by powers of ten. This has obvious advantages over the arbitrary set of conversions between units in the English system. The multiplying factors used in the SI system are expressed as a prefix to the name of the unit. These prefixes are as follows:

giga	10^9 (one billion)
mega	10^6 (one million)
kilo	10^3 (one thousand)
hecto	10^2 (one hundred)
deca	10^1 (ten)
deci	10^{-1} (one-tenth)
centi	10^{-2} (one-hundredth)
milli	10^{-3} (one-thousandth)
micro	10^{-6} (one-millionth)
nano	10^{-9} (one-billionth)

Thus, a kilometer is equal to one thousand meters, and a micrometer (often called simply a micron) is equal to one-millionth of a meter.

The following is a set of conversion factors between a few common units in the English and SI systems of units. These are multiplying factors, and are to be used for conversions in the direction indicated by the arrows; i.e.,

inch (2.540 →) centimeter
(← 0.3937)

indicates that a length in inches should be multiplied by 2.540 to determine the length in centimeters, and a length in centimeters should be multiplied by 0.3937 to determine the length in inches.

Length:

    inch (2.540 →) centimeter
         (← 0.3937)
         (25.40 →) millimeter
         (← 0.0394)

    foot (30.48 →) centimeter
         (← 0.0328)
         (0.3048 →) meter
         (← 3.2808)

    yard (0.9114 →) meter
         (← 1.0936)

    mile (1.609 →) kilometer
         (← 0.6215)

    mil (25.40 →) micrometer (micron)
         (← 0.0394)

Volume:

    quart (0.9463 →) liter (10^3 cubic centimeters)
         (← 1.0567)

    gallon (3.785 →) liter
         (← 0.2642)

Mass/Force:

    pound (mass) (0.4536 →   ) kilogram
                 (← 2.2046   )
                 (453.6 →   ) gram
                 (← 2.205×10^{-3})
    pound (force) (4.448 →   ) newton
                 (← 0.2248   )

    ounce   (28.35 →   ) gram
                 (← 0.0353   )
    ton   (907.2 →   ) kilogram
                 (← 1.102×10^{-3})
                 (0.9072 →   ) metric ton
                 (← 1.1023   )

Pressure (Stress):

psi (6.90×10^3 →) megapascal
(← 144.9)
(6.90×10^3 →) pascal
(← 1.449×10^{-4})
(7.03×10^{-4} →) kg/mm²
(← 1422)

Energy/Power:

ft-lb (1.356 →) joule
(← 0.7375)
Btu (1054 →) joule
(← 9.488×10^{-4})
(252.0 →) calorie*
(← 3.97×10^{-3})
Btu/hr (0.293 →) watt
(← 3.413)
horsepower (745.8 →) watt
(← 1.341×10^{-3})

* The calorie is not a standard SI energy unit.

TABLE A.1. Ceramic raw materials and formulas

Mineral Name	Mineral Formula	Formula Weight *(lb)*
Albite (soda spar)	$Na_2O \cdot Al_2O_3 \cdot 6SiO_2$	525.1
Alumina	Al_2O_3	101.9
Anatase (see titania)	. . .	. . .
Andalusite	Al_2SiO_5	162.3
Anhydrite	$CaSO_4$	136.2
Anorthite	$CaO \cdot Al_2O_3 \cdot 2SiO_2$	278.7
Antimony oxide	Sb_2O_3	291.5
Aragonite (see calcium carbonate)	. . .	. . .
Arsenious oxide	As_2O_3	197.8
Barium carbonate	$BaCO_3$	197.4
Barium chloride	$BaCl_2 \cdot 2H_2O$	244.3
Barium chromate	$BaCrO_4$	253.5
Barium hydroxide	$Ba(OH)_2 \cdot 8H_2O$	315.5
Barium oxide	BaO	153.4
Barium sulfate (barite)	$BaSO_4$	233.4
Bismuth oxide	Bi_2O_3	466.0
Bone ash	$13CaO \cdot 4P_2O_5 \cdot CO_2$ (approx.)	1,340.0
Borax	$Na_2B_4O_7 \cdot 10H_2O$	381.4
Boric acid	H_3BO_3	61.8
Boric oxide	B_2O_3	69.6
Calcite (see calcium carbonate)	. . .	. . .
Calcium borate (colemanite)	$Ca(BO_2)_2 \cdot 2H_2O$	161.7
Calcium carbonate (whiting)	$CaCO_3$	100.1
Calcium chloride	$CaCl_2 \cdot 6H_2O$	219.1
Calcium chloride (anhydrous)	$CaCl_2$	111.0
Calcium fluoride (fluorspar)	CaF_2	78.1
Calcium hydroxide	$Ca(OH)_2$	74.1
Calcium orthophosphate	$Ca_3(PO_4)_2$	310.3
Calcium oxide (lime)	CaO	56.1
Calcium sulfate (gypsum)	$CaSO_4 \cdot 2H_2O$	172.2
Carbon dioxide	CO_2	44.0
Chromium oxide	Cr_2O_3	152.0
Clay (kaolinite, china clay)	$Al_2Si_2O_5(OH)_4$	258.1
Cobaltic chloride	$CoCl_3$	165.3
Cobalt II, III opide	Co_3O_4	240.8
Cobalt III oxide	Co_2O_3	165.9
Cobaltous acetate	$Co(CH_3COO)_2 \cdot 4H_2O$	249.0
Cobaltous carbonate	$CoCO_3$	119.0
Cobaltous chloride	$CoCl_2 \cdot 6H_2O$	238.0
Cobaltous nitrate	$Co(NO_3)_2 \cdot 6H_2O$	291.1
Cobaltous oxide	CoO	74.9
Cobaltous phosphate	$Co_3(PO_4)_2 \cdot 3H_2O$	420.9
Cordierite	$Mg_2Al_4Si_5O_{18}$	584.9
Corundum (see alumina)	. . .	. . .
Cryolite	Na_3AlF_6	285.0
Cupric carbonate (basic)	$CuCO_3 \cdot Cu(OH)_2$	221.2
Cupric chloride	$CuCl_2 \cdot 2H_2O$	170.5
Cupric hydroxide	$Cu(OH)_2$	97.6
Cupric nitrate	$Cu(NO_3)_2 \cdot 6H_2O$	295.7
Cupric oxide	CuO	79.6
Cupric sulfate	$CuSO_4 \cdot 5H_2O$	249.7
Cuprous chloride	$CuCl$	99.0
Cuprous hydroxide	$Cu(OH)$	80.6
Cuprous oxide	Cu_2O	143.1
Cuprous sulfate	$Cu_2SO_4 \cdot H_2O$	225.2

TABLE A.1. **(Continued)**

Mineral Name	Mineral Formula	Formula Weight *(lb)*
Diopside	$CaSiO_3 \cdot MgSiO_3$	216.6
Dolomite	$CaCO_3 \cdot MgCO_3$	184.4
Feldspar (see albite, anorthite, orthoclase)	...	...
Ferric chloride	$FeCl_3$	162.2
Ferric hydroxide	$Fe(OH)_3$	106.9
Ferric oxide (hematite)	Fe_2O_3	159.7
Ferric sulfate	$Fe_2(SO_4)_3 \cdot 9H_2O$	562.0
Ferro-ferric oxide (magnetite)	Fe_3O_4	231.5
Ferrous carbonate (siderite)	$FeCO_3$	115.8
Ferrous oxide (wustite)	FeO	71.8
Ferrous sulfate	$FeSO_4 \cdot 7H_2O$	278.0
Ferrous sufide	FeS	87.9
Flint (see silica)	...	...
Gypsum (see calcium sulfate)	...	...
Ilmenite	$FeTiO_3$	151.9
Kaolinite (see clay)	...	...
Kyanite	Al_2SiO_5	162.3
Lead borate	$Pb(BO_2)_2 \cdot H_2O$	310.9
Lead carbonate	$PbCO_3$	267.2
Lead carbonate basic (white lead)	$2PbCO_3 \cdot Pb(OH)_2$	775.6
Lead chloride	$PbCl_2$	278.1
Lead dioxide	PbO_2	239.2
Lead oxide (litharge)	PbO	223.2
Lead oxide (red lead)	Pb_3O_4	685.6
Lithium carbonate	Li_2CO_3	73.9
Magnesium carbonate (magnesite)	$MgCO_3$	84.3
Magnesium chloride	$MgCl_2 \cdot 6H_2O$	203.3
Magnesium oxide (magnesia, periclase)	MgO	40.3
Manganese dioxide	MnO_2	86.9
Manganous carbonate	$MnCO_3$	114.9
Manganous oxide	MnO	70.9
Microcline (see orthoclase)	...	...
Mullite	$Al_6Si_2O_{13}$	425.9
Nickel chloride	$NiCl_2$	129.6
Nickel oxide	NiO	74.7
Niter (saltpeter) (see potassium nitrate)	...	...
Orthoclase (potash spar)	$K_2O \cdot Al_2O_3 \cdot 6SiO_2$	556.8
Potash spar (see orthoclase)	...	...
Potassium carbonate	K_2CO_3	138.0
Potassium chloride	KCl	74.6
Potassium chromate	K_2CrO_4	194.2
Potassium dichromate	$K_2Cr_2O_7$	294.2
Potassium ferrocyanide	$K_4Fe(CN)_6 \cdot 3H_2O$	422.4
Potassium hydroxide	KOH	56.1
Potassium nitrate (niter)	KNO_3	101.1
Potassium oxide (potash)	K_2O	94.2
Potassium permanganate	$KMnO_4$	158.0
Pyrophyllite	$Al_2Si_4O_{10}(OH)_2$	360.3
Quartz (see silica)	...	...

TABLE A.1. **(Continued)**

Mineral Name	Mineral Formula	Formula Weight (lb)
Silica (quartz, flint)	SiO_2	60.1
Silicic acid	H_2SiO_3	78.1
Sillimanite	Al_2SiO_5	162.3
Soda ash (see sodium carbonate)	. . .	. . .
Soda spar (see albite)	. . .	. . .
Sodium bicarbonate	$NaHCO_3$	84.0
Sodium carbonate (anhydrous)	Na_2CO_3	106.0
Sodium carbonate (hydrated) (soda ash)	$Na_2CO_3 \cdot 10H_2O$	286.2
Sodium chloride (salt)	$NaCl$	58.4
Sodium chromate	$Na_2CrO_4 \cdot 10H_2O$	342.2
Sodium dichromate	$Na_2Cr_2O_7 \cdot 2H_2O$	298.0
Sodium hydroxide (caustic, lye)	$NaOH$	40.0
Sodium nitrate (soda niter)	$NaNO_3$	85.0
Sodium oxide (soda)	Na_2O	62.0
Sodium silicate	variable $Na_2O{:}SiO_2$ ratios	. . .
Sodium sulfate (salt cake)	$Na_2SO_4 \cdot 10H_2O$	322.2
Spinel	$MgAl_2O_4$	142.2
Strontium carbonate	$SrCO_3$	147.6
Strontium oxide	SrO	103.6
Sulfur dioxide	SO_2	64.1
Sulfur trioxide	SO_3	80.1
Talc	$Mg_3Si_4O_{10}(OH)_2$	379.3
Tin chloride (stannic)	$SnCl_4$	260.5
Tin chloride (stannous)	$SnCl_2$	189.6
Tin oxide (stannic)	SnO_2	150.7
Tin oxide (stannous)	SnO	134.7
Titania (rutile, anatase)	TiO_2	80.1
Uranium dioxide	UO_2	270.0
Uranium oxide	U_3O_8	842.0
Uranium trioxide	UO_3	286.0
Wollastonite	$CaSiO_3$	116.2
Zinc carbonate	$ZnCO_3$	125.4
Zinc oxide	ZnO	81.4
Zinc sulfate	$ZnSO_4 \cdot 7H_2O$	287.6
Zirconia	ZrO_2	123.0
Zirconium silicate (zircon)	$ZrSiO_4$	183.3

TABLE A.2. Temperature conversion table

100 to 1250

C.	F.		C.	F.	
38	100	212	332	630	1166
43	110	230	338	640	1184
49	120	248	343	650	1202
54	130	266	349	660	1220
60	140	284	354	670	1238
66	150	302	360	680	1256
71	160	320	366	690	1274
77	170	338	371	700	1292
82	180	356	377	710	1310
88	190	374	382	720	1328
93	200	392	388	730	1346
99	210	410	393	740	1364
100	212	413	399	750	1382
104	220	428	404	760	1400
110	230	446	410	770	1418
116	240	464	416	780	1436
121	250	482	421	790	1454
127	260	500	427	800	1472
132	270	518	432	810	1490
138	280	536	438	820	1508
143	290	554	443	830	1526
149	300	572	449	840	1544
154	310	590	454	850	1562
160	320	608	460	860	1580
166	330	626	466	870	1598
171	340	644	471	880	1616
177	350	662	477	890	1634
182	360	680	482	900	1652
188	370	698	488	910	1670
193	380	716	493	920	1688
199	390	734	499	930	1706
204	400	752	504	940	1724
210	410	770	510	950	1742
216	420	788	516	960	1760
221	430	806	521	970	1778
227	440	824	527	980	1796
232	450	842	532	990	1814
238	460	860	538	1000	1832
243	470	878	543	1010	1850
249	480	896	549	1020	1868
254	490	914	554	1030	1886
260	500	932	560	1040	1904
266	510	950	566	1050	1922
271	520	968	571	1060	1940
277	530	986	577	1070	1958
282	540	1004	582	1080	1976
288	550	1022	588	1090	1994
293	560	1040	593	1100	2012
299	570	1058	599	1110	2030
304	580	1076	604	1120	2048
310	590	1094	610	1130	2066
316	600	1112	616	1140	2084
321	610	1130	621	1150	2102
327	620	1148	627	1160	2120
			632	1170	2138
			638	1180	2156
			643	1190	2174
			649	1200	2192
			654	1210	2210
			660	1220	2228
			666	1230	2246
			671	1240	2264
			677	1250	2282

INTERPOLATION FACTORS

C.		F.
0.56	1	1.8
1.11	2	3.6
1.67	3	5.4
2.22	4	7.2
2.78	5	9.0
3.33	6	10.8
3.89	7	12.6
4.44	8	14.4
5.00	9	16.2
5.56	10	18.0

1250 to 2500

C.	F.		C.	F.	
677	1250	2282	1027	1880	3416
682	1260	2300	1032	1890	3434
688	1270	2318	1038	1900	3452
693	1280	2336	1043	1910	3470
699	1290	2354	1049	1920	3488
704	1300	2372	1054	1930	3506
710	1310	2390	1060	1940	3524
716	1320	2408	1066	1950	3542
721	1330	2426	1071	1960	3560
727	1340	2444	1077	1970	3578
732	1350	2462	1082	1980	3596
738	1360	2480	1088	1990	3614
743	1370	2498	1093	2000	3632
749	1380	2516	1099	2010	3650
754	1390	2534	1104	2020	3668
760	1400	2552	1110	2030	3686
766	1410	2570	1116	2040	3704
771	1420	2588	1121	2050	3722
777	1430	2606	1127	2060	3740
782	1440	2624	1132	2070	3758
788	1450	2642	1138	2080	3776
793	1460	2660	1143	2090	3794
799	1470	2678	1149	2100	3812
804	1480	2696	1154	2110	3830
810	1490	2714	1160	2120	3848
816	1500	2732	1166	2130	3866
821	1510	2750	1171	2140	3884
827	1520	2768	1177	2150	3902
832	1530	2786	1182	2160	3920
838	1540	2804	1188	2170	3938
843	1550	2822	1193	2180	3956
849	1560	2840	1199	2190	3974
854	1570	2858	1204	2200	3992
860	1580	2876	1210	2210	4010
866	1590	2894	1216	2220	4028
871	1600	2912	1221	2230	4046
877	1610	2930	1227	2240	4064
882	1620	2948	1232	2250	4082
888	1630	2966	1238	2260	4100
893	1640	2984	1243	2270	4118
899	1650	3002	1249	2280	4136
904	1660	3020	1254	2290	4154
910	1670	3038	1260	2300	4172
916	1680	3056	1266	2310	4190
921	1690	3074	1271	2320	4208
927	1700	3092	1277	2330	4226
932	1710	3110	1282	2340	4244
938	1720	3128	1288	2350	4262
943	1730	3146	1293	2360	4280
949	1740	3164	1299	2370	4298
954	1750	3182	1304	2380	4316
960	1760	3200	1310	2390	4334
966	1770	3218	1316	2400	4352
971	1780	3236	1321	2410	4370
977	1790	3254	1327	2420	4388
982	1800	3272	1332	2430	4406
988	1810	3290	1338	2440	4424
993	1820	3308	1343	2450	4442
999	1830	3326	1349	2460	4460
1004	1840	3344	1354	2470	4478
1010	1850	3362	1360	2480	4496
1016	1860	3380	1366	2490	4514
1021	1870	3398	1371	2500	4532

2500 to 3750

C.	F.		C.	F.	
1371	2500	4532	1721	3130	5666
1377	2510	4550	1727	3140	5684
1382	2520	4568	1732	3150	5702
1388	2530	4586	1738	3160	5720
1393	2540	4604	1743	3170	5738
1399	2550	4622	1749	3180	5756
1404	2560	4640	1754	3190	5774
1410	2570	4658	1760	3200	5792
1416	2580	4676	1766	3210	5810
1421	2590	4694	1771	3220	5828
1427	2600	4712	1777	3230	5846
1432	2610	4730	1782	3240	5864
1438	2620	4748	1788	3250	5882
1443	2630	4766	1793	3260	5900
1449	2640	4784	1799	3270	5918
1454	2650	4802	1804	3280	5936
1460	2660	4820	1810	3290	5954
1466	2670	4838	1816	3300	5972
1471	2680	4856	1821	3310	5990
1477	2690	4874	1827	3320	6008
1482	2700	4892	1832	3330	6026
1488	2710	4910	1838	3340	6044
1493	2720	4928	1843	3350	6062
1499	2730	4946	1849	3360	6080
1504	2740	4964	1854	3370	6098
1510	2750	4982	1860	3380	6116
1516	2760	5000	1866	3390	6134
1521	2770	5018	1871	3400	6152
1527	2780	5036	1877	3410	6170
1532	2790	5054	1882	3420	6188
1538	2800	5072	1888	3430	6206
1543	2810	5090	1893	3440	6224
1549	2820	5108	1899	3450	6242
1554	2830	5126	1904	3460	6260
1560	2840	5144	1910	3470	6278
1566	2850	5162	1916	3480	6296
1571	2860	5180	1921	3490	6314
1577	2870	5198	1927	3500	6332
1582	2880	5216	1932	3510	6350
1588	2890	5234	1938	3520	6368
1593	2900	5252	1943	3530	6386
1599	2910	5270	1949	3540	6404
1604	2920	5288	1954	3550	6422
1610	2930	5306	1960	3560	6440
1616	2940	5324	1966	3570	6458
1621	2950	5342	1971	3580	6476
1627	2960	5360	1977	3590	6494
1632	2970	5378	1982	3600	6512
1638	2980	5396	1988	3610	6530
1643	2990	5414	1993	3620	6548
1649	3000	5432	1999	3630	6566
1654	3010	5450	2004	3640	6584
1660	3020	5468	2010	3650	6602
1666	3030	5486	2016	3660	6620
1671	3040	5504	2021	3670	6638
1677	3050	5522	2027	3680	6656
1682	3060	5540	2032	3690	6674
1688	3070	5558	2038	3700	6692
1693	3080	5576	2043	3710	6710
1699	3090	5594	2049	3720	6728
1704	3100	5612	2054	3730	6746
1710	3110	5630	2060	3740	6764
1716	3120	5648	2066	3750	6782

3750 to 5000

C.	F.		C.	F.	
2066	3750	6782	2416	4380	7916
2071	3760	6800	2421	4390	7934
2077	3770	6818	2427	4400	7952
2082	3780	6836	2432	4410	7970
2088	3790	6854	2438	4420	7988
2093	3800	6872	2443	4430	8006
2099	3810	6890	2449	4440	8024
2104	3820	6908	2454	4450	8042
2110	3830	6926	2460	4460	8060
2116	3840	6944	2466	4470	8078
2121	3850	6962	2471	4480	8096
2127	3860	6980	2477	4490	8114
2132	3870	6998	2482	4500	8132
2138	3880	7016	2488	4510	8150
2143	3890	7034	2493	4520	8168
2149	3900	7052	2499	4530	8186
2154	3910	7070	2504	4540	8204
2160	3920	7088	2510	4550	8222
2166	3930	7106	2516	4560	8240
2171	3940	7124	2521	4570	8258
2177	3950	7142	2527	4580	8276
2182	3960	7160	2532	4590	8294
2188	3970	7178	2538	4600	8312
2193	3980	7196	2543	4610	8330
2199	3990	7214	2549	4620	8348
2204	4000	7232	2554	4630	8366
2210	4010	7250	2560	4640	8384
2216	4020	7268	2566	4650	8402
2221	4030	7286	2571	4660	8420
2227	4040	7304	2577	4670	8438
2232	4050	7322	2582	4680	8456
2238	4060	7340	2588	4690	8474
2243	4070	7358	2593	4700	8492
2249	4080	7376	2599	4710	8510
2254	4090	7394	2604	4720	8528
2260	4100	7412	2610	4730	8546
2266	4110	7430	2616	4740	8564
2271	4120	7448	2621	4750	8582
2277	4130	7466	2627	4760	8600
2282	4140	7484	2632	4770	8618
2288	4150	7502	2638	4780	8636
2293	4160	7520	2643	4790	8654
2299	4170	7538	2649	4800	8672
2304	4180	7556	2654	4810	8690
2310	4190	7574	2660	4820	8708
2316	4200	7592	2666	4830	8726
2321	4210	7610	2671	4840	8744
2327	4220	7628	2677	4850	8762
2332	4230	7646	2682	4860	8780
2338	4240	7664	2688	4870	8798
2343	4250	7682	2693	4880	8816
2349	4260	7700	2699	4890	8834
2354	4270	7718	2704	4900	8852
2360	4280	7736	2710	4910	8870
2366	4290	7754	2716	4920	8888
2371	4300	7772	2721	4930	8906
2377	4310	7790	2727	4940	8924
2382	4320	7808	2732	4950	8942
2388	4330	7826	2738	4960	8960
2393	4340	7844	2743	4970	8978
2399	4350	7862	2749	4980	8996
2404	4360	7880	2754	4990	9014
2410	4370	7898	2760	5000	9032

Source: Reprinted by permission of Corhart Refractories Co., Louisville, Kentucky.

TABLE A.3. Output of type J thermocouple

IRON vs. CONSTANTAN THERMOCOUPLE
Degrees Fahrenheit — Reference Junction 32° F.

°F	0	1	2	3	4	5	6	7	8	9
					Millivolts					
-310	-7.66	-7.68	-7.69	-7.70	-7.71	-7.73	-7.74	-7.75	-7.76	-7.78
-300	-7.52	-7.54	-7.55	-7.57	-7.58	-7.59	-7.61	-7.62	-7.64	-7.65
-290	-7.38	-7.39	-7.40	-7.42	-7.44	-7.45	-7.46	-7.48	-7.49	-7.51
-280	-7.22	-7.24	-7.25	-7.27	-7.28	-7.30	-7.31	-7.33	-7.34	-7.36
-270	-7.06	-7.07	-7.09	-7.11	-7.12	-7.14	-7.15	-7.17	-7.19	-7.20
-260	-6.89	-6.90	-6.92	-6.94	-6.96	-6.97	-6.99	-7.01	-7.02	-7.04
-250	-6.71	-6.73	-6.75	-6.77	-6.78	-6.80	-6.82	-6.84	-6.85	-6.87
-240	-6.53	-6.55	-6.57	-6.59	-6.61	-6.62	-6.64	-6.66	-6.68	-6.70
-230	-6.35	-6.37	-6.38	-6.40	-6.42	-6.44	-6.46	-6.48	-6.50	-6.52
-220	-6.16	-6.18	-6.19	-6.21	-6.23	-6.25	-6.27	-6.29	-6.31	-6.33
-210	-5.96	-5.98	-6.00	-6.02	-6.04	-6.06	-6.08	-6.10	-6.12	-6.14
-200	-5.76	-5.78	-5.80	-5.82	-5.84	-5.86	-5.88	-5.90	-5.92	-5.94
-190	-5.55	-5.57	-5.59	-5.61	-5.63	-5.65	-5.67	-5.70	-5.72	-5.74
-180	-5.34	-5.36	-5.38	-5.40	-5.42	-5.44	-5.46	-5.49	-5.51	-5.53
-170	-5.12	-5.14	-5.16	-5.19	-5.21	-5.23	-5.25	-5.27	-5.30	-5.32
-160	-4.90	-4.92	-4.94	-4.97	-4.99	-5.01	-5.03	-5.06	-5.08	-5.10
-150	-4.68	-4.70	-4.72	-4.74	-4.76	-4.79	-4.81	-4.83	-4.86	-4.88
-140	-4.44	-4.47	-4.49	-4.51	-4.54	-4.56	-4.58	-4.61	-4.63	-4.65
-130	-4.21	-4.23	-4.26	-4.28	-4.30	-4.33	-4.35	-4.38	-4.40	-4.42
-120	-3.97	-4.00	-4.02	-4.04	-4.07	-4.09	-4.12	-4.14	-4.16	-4.19
-110	-3.73	-3.76	-3.78	-3.81	-3.83	-3.85	-3.88	-3.90	-3.93	-3.95
-100	-3.49	-3.51	-3.54	-3.56	-3.59	-3.61	-3.64	-3.66	-3.68	-3.71
-90	-3.24	-3.27	-3.29	-3.32	-3.34	-3.36	-3.39	-3.41	-3.44	-3.46
-80	-2.99	-3.02	-3.04	-3.07	-3.09	-3.12	-3.14	-3.17	-3.19	-3.22
-70	-2.74	-2.76	-2.79	-2.81	-2.84	-2.86	-2.89	-2.92	-2.94	-2.97
-60	-2.48	-2.51	-2.53	-2.56	-2.58	-2.61	-2.64	-2.66	-2.69	-2.71
-50	-2.22	-2.25	-2.27	-2.30	-2.33	-2.35	-2.38	-2.40	-2.43	-2.46
-40	-1.96	-1.99	-2.01	-2.04	-2.06	-2.09	-2.12	-2.14	-2.17	-2.20
-30	-1.70	-1.72	-1.75	-1.78	-1.80	-1.83	-1.86	-1.88	-1.91	-1.94
-20	-1.43	-1.46	-1.48	-1.51	-1.54	-1.56	-1.59	-1.62	-1.64	-1.67
-10	-1.16	-1.19	-1.21	-1.24	-1.27	-1.29	-1.32	-1.35	-1.38	-1.40
(-)0	-.89	-.91	-.94	-.97	-1.00	-1.02	-1.05	-1.08	-1.10	-1.13
(+)0	-.89	-.86	-.83	-.80	-.78	-.75	-.72	-.70	-.67	-.64
10	-.61	-.58	-.56	-.53	-.50	-.48	-.45	-.42	-.39	-.36
20	-.34	-.31	-.28	-.25	-.22	-.20	-.17	-.14	-.11	-.09
30	-.06	-.03	.00	.03	.05	.08	.11	.14	.17	.19
40	.22	.25	.28	.31	.34	.36	.39	.42	.45	.48
50	.50	.53	.56	.59	.62	.65	.67	.70	.73	.76
60	.79	.82	.84	.87	.90	.93	.96	.99	1.02	1.04
70	1.07	1.10	1.13	1.16	1.19	1.22	1.25	1.28	1.30	1.33
80	1.36	1.39	1.42	1.45	1.48	1.51	1.54	1.56	1.59	1.62
90	1.65	1.68	1.71	1.74	1.77	1.80	1.83	1.85	1.88	1.91
100	1.94	1.97	2.00	2.03	2.06	2.09	2.12	2.14	2.17	2.20
110	2.23	2.26	2.29	2.32	2.35	2.38	2.41	2.44	2.47	2.50
120	2.52	2.55	2.58	2.61	2.64	2.67	2.70	2.73	2.76	2.79
130	2.82	2.85	2.88	2.91	2.94	2.97	3.00	3.03	3.06	3.08
140	3.11	3.14	3.17	3.20	3.23	3.26	3.29	3.32	3.35	3.38
150	3.41	3.44	3.47	3.50	3.53	3.56	3.59	3.62	3.65	3.68
160	3.71	3.74	3.77	3.80	3.83	3.86	3.89	3.92	3.95	3.98
170	4.01	4.04	4.07	4.10	4.13	4.16	4.19	4.22	4.25	4.28
180	4.31	4.34	4.37	4.40	4.43	4.46	4.49	4.52	4.55	4.58
190	4.61	4.64	4.67	4.70	4.73	4.76	4.79	4.82	4.85	4.88
200	4.91	4.94	4.97	5.00	5.03	5.06	5.09	5.12	5.15	5.18
210	5.21	5.24	5.27	5.30	5.33	5.36	5.39	5.42	5.45	5.48
220	5.51	5.54	5.57	5.60	5.63	5.66	5.69	5.72	5.75	5.78
230	5.81	5.84	5.87	5.90	5.93	5.96	5.99	6.02	6.05	6.08
240	6.11	6.14	6.17	6.20	6.24	6.27	6.30	6.33	6.36	6.39
250	6.42	6.45	6.48	6.51	6.54	6.57	6.60	6.63	6.66	6.69
260	6.72	6.75	6.78	6.81	6.84	6.87	6.90	6.93	6.96	7.00
270	7.03	7.06	7.09	7.12	7.15	7.18	7.21	7.24	7.27	7.30
280	7.33	7.36	7.39	7.42	7.45	7.48	7.51	7.54	7.58	7.61
290	7.64	7.67	7.70	7.73	7.76	7.79	7.82	7.85	7.88	7.91

TABLE A.3. (Continued)

IRON vs. CONSTANTAN THERMOCOUPLE
Degrees Fahrenheit   Reference Junction 32° F.

°F	0	1	2	3	4	5	6	7	8	9
					Millivolts					
300	7.94	7.97	8.00	8.04	8.07	8.10	8.13	8.16	8.19	8.22
310	8.25	8.28	8.31	8.34	8.37	8.40	8.44	8.47	8.50	8.53
320	8.56	8.59	8.62	8.65	8.68	8.71	8.74	8.77	8.80	8.84
330	8.87	8.90	8.93	8.96	8.99	9.02	9.05	9.08	9.11	9.14
340	9.17	9.20	9.24	9.27	9.30	9.33	9.36	9.39	9.42	9.45
350	9.48	9.51	9.54	9.58	9.61	9.64	9.67	9.70	9.73	9.76
360	9.79	9.82	9.85	9.88	9.92	9.95	9.98	10.01	10.04	10.07
370	10.10	10.13	10.16	10.19	10.22	10.25	10.28	10.32	10.35	10.38
380	10.41	10.44	10.47	10.50	10.53	10.56	10.60	10.63	10.66	10.69
390	10.72	10.75	10.78	10.81	10.84	10.87	10.90	10.94	10.97	11.00
400	11.03	11.06	11.09	11.12	11.15	11.18	11.21	11.24	11.28	11.31
410	11.34	11.37	11.40	11.43	11.46	11.49	11.52	11.55	11.58	11.62
420	11.65	11.68	11.71	11.74	11.77	11.80	11.83	11.86	11.89	11.92
430	11.96	11.99	12.02	12.05	12.08	12.11	12.14	12.17	12.20	12.23
440	12.26	12.30	12.33	12.36	12.39	12.42	12.45	12.48	12.51	12.54
450	12.57	12.60	12.64	12.67	12.70	12.73	12.76	12.79	12.82	12.85
460	12.88	12.91	12.94	12.98	13.01	13.04	13.07	13.10	13.13	13.16
470	13.19	13.22	13.25	13.28	13.31	13.34	13.38	13.41	13.44	13.47
480	13.50	13.53	13.56	13.59	13.62	13.65	13.68	13.72	13.75	13.78
490	13.81	13.84	13.87	13.90	13.93	13.96	13.99	14.02	14.05	14.08
500	14.12	14.15	14.18	14.21	14.24	14.27	14.30	14.33	14.36	14.39
510	14.42	14.45	14.48	14.52	14.55	14.58	14.61	14.64	14.67	14.70
520	14.73	14.76	14.79	14.82	14.85	14.88	14.91	14.94	14.98	15.01
530	15.04	15.07	15.10	15.13	15.16	15.19	15.22	15.25	15.28	15.31
540	15.34	15.37	15.40	15.44	15.47	15.50	15.53	15.56	15.59	15.62
550	15.65	15.68	15.71	15.74	15.77	15.80	15.84	15.87	15.90	15.93
560	15.96	15.99	16.02	16.05	16.08	16.11	16.14	16.17	16.20	16.23
570	16.26	16.30	16.33	16.36	16.39	16.42	16.45	16.48	16.51	16.54
580	16.57	16.60	16.63	16.66	16.69	16.72	16.75	16.78	16.82	16.85
590	16.88	16.91	16.94	16.97	17.00	17.03	17.06	17.09	17.12	17.15
600	17.18	17.21	17.24	17.28	17.31	17.34	17.37	17.40	17.43	17.46
610	17.49	17.52	17.55	17.58	17.61	17.64	17.68	17.71	17.74	17.77
620	17.80	17.83	17.86	17.89	17.92	17.95	17.98	18.01	18.04	18.08
630	18.11	18.14	18.17	18.20	18.23	18.26	18.29	18.32	18.35	18.38
640	18.41	18.44	18.47	18.50	18.54	18.57	18.60	18.63	18.66	18.69
650	18.72	18.75	18.78	18.81	18.84	18.87	18.90	18.94	18.97	19.00
660	19.03	19.06	19.09	19.12	19.15	19.18	19.21	19.24	19.27	19.30
670	19.34	19.37	19.40	19.43	19.46	19.49	19.52	19.55	19.58	19.61
680	19.64	19.67	19.70	19.74	19.77	19.80	19.83	19.86	19.89	19.92
690	19.95	19.98	20.01	20.04	20.07	20.10	20.13	20.16	20.20	20.23
700	20.26	20.29	20.32	20.35	20.38	20.41	20.44	20.47	20.50	20.53
710	20.56	20.59	20.62	20.66	20.69	20.72	20.75	20.78	20.81	20.84
720	20.87	20.90	20.93	20.96	20.99	21.02	21.05	21.08	21.11	21.14
730	21.18	21.21	21.24	21.27	21.30	21.33	21.36	21.39	21.42	21.45
740	21.48	21.51	21.54	21.57	21.60	21.64	21.67	21.70	21.73	21.76
750	21.79	21.82	21.85	21.88	21.91	21.94	21.97	22.00	22.03	22.06
760	22.10	22.13	22.16	22.19	22.22	22.25	22.28	22.31	22.34	22.37
770	22.40	22.43	22.46	22.49	22.52	22.55	22.58	22.62	22.65	22.68
780	22.71	22.74	22.77	22.80	22.83	22.86	22.89	22.92	22.95	22.98
790	23.01	23.04	23.08	23.11	23.14	23.17	23.20	23.23	23.26	23.29
800	23.32	23.35	23.38	23.41	23.44	23.47	23.50	23.53	23.56	23.60
810	23.63	23.66	23.69	23.72	23.75	23.78	23.81	23.84	23.87	23.90
820	23.93	23.96	23.99	24.02	24.06	24.09	24.12	24.15	24.18	24.21
830	24.24	24.27	24.30	24.33	24.36	24.39	24.42	24.45	24.48	24.52
840	24.55	24.58	24.61	24.64	24.67	24.70	24.73	24.76	24.79	24.82
850	24.85	24.88	24.91	24.94	24.98	25.01	25.04	25.07	25.10	25.13
860	25.16	25.19	25.22	25.25	25.28	25.32	25.35	25.38	25.41	25.44
870	25.47	25.50	25.53	25.56	25.59	25.62	25.65	25.68	25.72	25.75
980	25.78	25.81	25.84	25.87	25.90	25.93	25.96	25.99	26.02	26.06
890	26.09	26.12	26.15	26.18	26.21	26.24	26.27	26.30	26.33	26.36

TABLE A.3. (Continued)

IRON vs. CONSTANTAN THERMOCOUPLE
Degrees Fahrenheit   Reference Junction 32° F.

°F	0	1	2	3	4	5	6	7	8	9
					Millivolts					
900	26.40	26.43	26.46	26.49	26.52	26.55	26.58	26.61	26.64	26.67
910	26.70	26.74	26.77	26.80	26.83	26.86	26.89	26.92	26.95	26.98
920	27.02	27.05	27.08	27.11	27.14	27.17	27.20	27.23	27.26	27.30
930	27.33	27.36	27.39	27.42	27.45	27.48	27.51	27.54	27.58	27.61
940	27.64	27.67	27.70	27.73	27.76	27.80	27.83	27.86	27.89	27.92
950	27.95	27.98	28.02	28.05	28.08	28.11	28.14	28.17	28.20	28.23
960	28.26	28.30	28.33	28.36	28.39	28.42	28.45	28.48	28.52	28.55
970	28.58	28.61	28.64	28.67	28.70	28.74	28.77	28.80	28.83	28.86
980	28.89	28.92	28.96	28.99	29.02	29.05	29.08	29.11	29.14	29.18
990	29.21	29.24	29.27	29.30	29.33	29.37	29.40	29.43	29.46	29.49
1000	29.52	29.56	29.59	29.62	29.65	29.68	29.71	29.75	29.78	29.81
1010	29.84	29.87	29.90	29.94	29.97	30.00	30.03	30.06	30.10	30.13
1020	30.16	30.19	30.22	30.25	30.28	30.32	30.35	30.38	30.41	30.44
1030	30.48	30.51	30.54	30.57	30.60	30.64	30.67	30.70	30.73	30.76
1040	30.80	30.83	30.86	30.89	30.92	30.96	30.99	31.02	31.05	31.08
1050	31.12	31.15	31.18	31.21	31.24	31.28	31.31	31.34	31.37	31.40
1060	31.44	31.47	31.50	31.53	31.56	31.60	31.63	31.66	31.69	31.72
1070	31.76	31.79	31.82	31.85	31.88	31.92	31.95	31.98	32.01	32.05
1080	32.08	32.11	32.14	32.18	32.21	32.24	32.27	32.30	32.34	32.37
1090	32.40	32.43	32.47	32.50	32.53	32.56	32.60	32.63	32.66	32.69
1100	32.72	32.76	32.79	32.82	32.86	32.89	32.92	32.95	32.98	33.02
1110	33.05	33.08	33.11	33.15	33.18	33.21	33.24	33.28	33.31	33.34
1120	33.37	33.41	33.44	33.47	33.50	33.54	33.57	33.60	33.64	33.67
1130	33.70	33.73	33.76	33.80	33.83	33.86	33.90	33.93	33.96	33.99
1140	34.03	34.06	34.09	34.12	34.16	34.19	34.22	34.26	34.29	34.32
1150	34.36	34.39	34.42	34.45	34.49	34.52	34.55	34.58	34.62	34.65
1160	34.68	34.72	34.75	34.78	34.82	34.85	34.88	34.92	34.95	34.98
1170	35.01	35.05	35.08	35.11	35.15	35.18	35.21	35.25	35.28	35.31
1180	35.35	35.38	35.41	35.45	35.48	35.51	35.54	35.58	35.61	35.64
1190	35.68	35.71	35.74	35.78	35.81	35.84	35.88	35.91	35.94	35.98
1200	36.01	36.05	36.08	36.11	36.15	36.18	36.21	36.25	36.28	36.31
1210	36.35	36.38	36.42	36.45	36.48	36.52	36.55	36.58	36.62	36.65
1220	36.69	36.72	36.75	36.79	36.82	36.86	36.89	36.92	36.96	36.99
1230	37.02	37.06	37.09	37.13	37.16	37.20	37.23	37.26	37.30	37.33
1240	37.36	37.40	37.43	37.47	37.50	37.54	37.57	37.60	37.64	37.67
1250	37.71	37.74	37.78	37.81	37.84	37.88	37.91	37.95	37.98	38.02
1260	38.05	38.08	38.12	38.15	38.19	38.22	38.26	38.29	38.32	38.36
1270	38.39	38.43	38.46	38.50	38.53	38.57	38.60	38.64	38.67	38.70
1280	38.74	38.77	38.81	38.84	38.88	38.91	38.95	38.98	39.02	39.05
1290	39.08	39.12	39.15	39.19	39.22	39.26	39.29	39.33	39.36	39.40
1300	39.43	39.47	39.50	39.54	39.57	39.61	39.64	39.68	39.71	39.75
1310	39.78	39.82	39.85	39.89	39.92	39.96	39.99	40.03	40.06	40.10
1320	40.13	40.17	40.20	40.24	40.27	40.31	40.34	40.38	40.41	40.45
1330	40.48	40.52	40.55	40.59	40.62	40.66	40.69	40.73	40.76	40.80
1340	40.83	40.87	40.90	40.94	40.98	41.01	41.05	41.08	41.12	41.15
1350	41.19	41.22	41.26	41.29	41.33	41.36	41.40	41.43	41.47	41.50
1360	41.54	41.58	41.61	41.65	41.68	41.72	41.75	41.79	41.82	41.86
1370	41.90	41.93	41.97	42.00	42.04	42.07	42.11	42.14	42.18	42.22
1380	42.25	42.29	42.32	42.36	42.39	42.43	42.46	42.50	42.53	42.57
1390	42.61	42.64	42.68	42.71	42.75	42.78	42.82	42.85	42.89	42.92
1400	42.96	43.00	43.03	43.07	43.10	43.14	43.18	43.21	43.25	43.28
1410	43.32	43.35	43.39	43.43	43.46	43.50	43.53	43.57	43.60	43.64
1420	43.68	43.71	43.75	43.78	43.82	43.85	43.89	43.92	43.96	44.00
1430	44.03	44.07	44.10	44.14	44.18	44.21	44.25	44.28	44.32	44.35
1440	44.39	44.42	44.46	44.50	44.53	44.57	44.60	44.64	44.68	44.71
1450	44.75	44.78	44.82	44.85	44.89	44.93	44.96	45.00	45.03	45.07
1460	45.10	45.14	45.18	45.21	45.25	45.28	45.32	45.35	45.39	45.42
1470	45.46	45.50	45.53	45.57	45.60	45.64	45.68	45.71	45.75	45.78
1480	45.82	45.85	45.89	45.92	45.96	46.00	46.03	46.07	46.10	46.14
1490	46.18	46.21	46.25	46.28	46.32	46.35	46.39	46.42	46.46	46.50

TABLE A.3. (Continued)

IRON vs. CONSTANTAN THERMOCOUPLE
Degrees Fahrenheit   Reference Junction 32° F.

°F	0	1	2	3	4	5	6	7	8	9
					Millivolts					
1500	46.53	46.57	46.60	46.64	46.67	46.71	46.74	46.78	46.82	46.85
1510	46.89	46.92	46.96	47.00	47.03	47.07	47.10	47.14	47.17	47.21
1520	47.24	47.28	47.32	47.35	47.39	47.42	47.46	47.49	47.53	47.56
1530	47.60	47.63	47.67	47.70	47.74	47.78	47.81	47.85	47.88	47.92
1540	47.95	47.99	48.02	48.06	48.09	48.13	48.16	48.20	48.24	48.27
1550	48.31	48.34	48.38	48.41	48.45	48.48	48.52	48.55	48.59	48.62
1560	48.66	48.69	48.73	48.76	48.80	48.83	48.87	48.90	48.94	48.97
1570	49.01	49.04	49.08	49.11	49.15	49.18	49.22	49.25	49.29	49.32
1580	49.36	49.39	49.43	49.46	49.50	49.53	49.56	49.60	49.63	49.67
1590	49.70	49.74	49.77	49.81	49.84	49.88	49.91	49.94	49.98	50.01
1600	50.05									

Source: National Bureau of Standards Circular 561, 1955. (Reprinted by permission of Leeds & Northrup Company.)

TABLE A.4. Output of type T thermocouple

COPPER vs. CONSTANTAN THERMOCOUPLE
Degrees Fahrenheit   Reference Junction 32° F.

°F	0	1	2	3	4	5	6	7	8	9
					Millivolts					
-310	-5.379	-5.388	-5.397	-5.406	....	....	....	....	....	....
-300	-5.284	-5.294	-5.303	-5.313	-5.322	-5.332	-5.341	-5.351	-5.360	-5.370
-290	-5.185	-5.195	-5.205	-5.215	-5.225	-5.235	-5.245	-5.254	-5.264	-5.274
-280	-5.081	-5.092	-5.102	-5.113	-5.124	-5.134	-5.144	-5.154	-5.165	-5.175
-270	-4.974	-4.985	-4.996	-5.007	-5.018	-5.029	-5.039	-5.050	-5.060	-5.071
-260	-4.863	-4.874	-4.885	-4.897	-4.908	-4.919	-4.930	-4.941	-4.952	-4.963
-250	-4.747	-4.759	-4.770	-4.782	-4.794	-4.805	-4.817	-4.829	-4.840	-4.851
-240	-4.627	-4.640	-4.652	-4.664	-4.676	-4.688	-4.700	-4.712	-4.724	-4.735
-230	-4.504	-4.517	-4.529	-4.542	-4.554	-4.566	-4.579	-4.591	-4.603	-4.615
-220	-4.377	-4.390	-4.403	-4.415	-4.428	-4.441	-4.454	-4.466	-4.479	-4.492
-210	-4.246	-4.259	-4.272	-4.286	-4.299	-4.312	-4.325	-4.338	-4.351	-4.364
-200	-4.111	-4.125	-4.138	-4.151	-4.165	-4.179	-4.192	-4.206	-4.219	-4.232
-190	-3.972	-3.986	-4.000	-4.014	-4.028	-4.042	-4.056	-4.069	-4.083	-4.097
-180	-3.829	-3.844	-3.858	-3.873	-3.887	-3.901	-3.915	-3.929	-3.944	-3.958
-170	-3.684	-3.698	-3.713	-3.727	-3.742	-3.757	-3.771	-3.786	-3.800	-3.815
-160	-3.533	-3.548	-3.564	-3.579	-3.594	-3.609	-3.624	-3.639	-3.654	-3.669
-150	-3.380	-3.396	-3.411	-3.426	-3.441	-3.457	-3.472	-3.488	-3.503	-3.518
-140	-3.223	-3.238	-3.254	-3.270	-3.286	-3.301	-3.317	-3.333	-3.349	-3.365
-130	-3.062	-3.078	-3.094	-3.110	-3.127	-3.143	-3.159	-3.175	-3.191	-3.207
-120	-2.897	-2.914	-2.931	-2.947	-2.964	-2.980	-2.997	-3.013	-3.030	-3.046
-110	-2.730	-2.747	-2.764	-2.781	-2.797	-2.814	-2.831	-2.847	-2.864	-2.881
-100	-2.559	-2.577	-2.594	-2.611	-2.628	-2.645	-2.662	-2.679	-2.696	-2.713
-90	-2.385	-2.402	-2.420	-2.437	-2.455	-2.472	-2.490	-2.507	-2.525	-2.542
-80	-2.207	-2.225	-2.243	-2.260	-2.278	-2.296	-2.314	-2.332	-2.349	-2.367
-70	-2.026	-2.044	-2.063	-2.081	-2.099	-2.117	-2.135	-2.153	-2.171	-2.189
-60	-1.842	-1.860	-1.879	-1.897	-1.916	-1.934	-1.953	-1.971	-1.989	-2.008
-50	-1.654	-1.673	-1.692	-1.711	-1.729	-1.748	-1.767	-1.786	-1.804	-1.823
-40	-1.463	-1.482	-1.502	-1.521	-1.540	-1.559	-1.578	-1.597	-1.616	-1.635
-30	-1.270	-1.289	-1.308	-1.328	-1.347	-1.367	-1.386	-1.406	-1.425	-1.444
-20	-1.072	-1.092	-1.112	-1.132	-1.152	-1.171	-1.191	-1.210	-1.230	-1.250
-10	-0.872	-0.893	-0.913	-0.933	-0.953	-0.973	-0.993	-1.013	-1.033	-1.053
(-)0	-0.670	-0.690	-0.710	-0.730	-0.751	-0.771	-0.792	-0.812	-0.832	-0.852
(+)0	-0.670	-0.649	-0.629	-0.608	-0.588	-0.567	-0.546	-0.526	-0.505	-0.484
10	-0.463	-0.442	-0.421	-0.401	-0.380	-0.359	-0.339	-0.318	-0.297	-0.275
20	-0.254	-0.233	-0.212	-0.191	-0.170	-0.149	-0.128	-0.107	-0.085	-0.064
30	-0.042	-0.021	0.000	0.021	0.042	0.064	0.086	0.107	0.129	0.150
40	0.171	0.193	0.215	0.236	0.258	0.280	0.302	0.324	0.346	0.367
50	0.389	0.411	0.433	0.455	0.477	0.499	0.521	0.543	0.565	0.587
60	0.609	0.631	0.654	0.676	0.698	0.720	0.743	0.765	0.787	0.809
70	0.832	0.854	0.877	0.899	0.922	0.944	0.967	0.990	1.012	1.035
80	1.057	1.080	1.103	1.126	1.148	1.171	1.194	1.217	1.240	1.263
90	1.286	1.309	1.332	1.355	1.378	1.401	1.424	1.448	1.471	1.494
100	1.517	1.540	1.563	1.587	1.610	1.633	1.657	1.680	1.704	1.727
110	1.751	1.774	1.798	1.821	1.845	1.869	1.893	1.916	1.940	1.963
120	1.987	2.011	2.035	2.059	2.083	2.107	2.131	2.154	2.178	2.202
130	2.226	2.250	2.274	2.298	2.322	2.346	2.370	2.394	2.418	2.443
140	2.467	2.491	2.516	2.540	2.565	2.589	2.614	2.638	2.663	2.687
150	2.711	2.736	2.760	2.785	2.810	2.835	2.859	2.884	2.908	2.933
160	2.958	2.982	3.007	3.032	3.057	3.082	3.107	3.132	3.157	3.182
170	3.207	3.232	3.257	3.282	3.307	3.332	3.357	3.382	3.407	3.433
180	3.458	3.483	3.508	3.534	3.559	3.584	3.610	3.635	3.661	3.686
190	3.712	3.737	3.762	3.787	3.813	3.839	3.864	3.890	3.915	3.941
200	3.967	3.993	4.018	4.044	4.070	4.096	4.122	4.148	4.174	4.199
210	4.225	4.251	4.277	4.303	4.329	4.355	4.381	4.408	4.434	4.460
220	4.486	4.512	4.538	4.564	4.590	4.617	4.643	4.670	4.696	4.722
230	4.749	4.775	4.801	4.827	4.854	4.880	4.907	4.934	4.960	4.987
240	5.014	5.040	5.067	5.094	5.120	5.147	5.174	5.200	5.227	5.254

TABLE A.4. (Continued)

COPPER vs. CONSTANTAN THERMOCOUPLE
Degrees Fahrenheit Reference Junction 32° F.

°F	0	1	2	3	4	5	6	7	8	9
					Millivolts					
250	5.280	5.307	5.334	5.361	5.388	5.415	5.442	5.469	5.496	5.523
260	5.550	5.577	5.604	5.631	5.658	5.685	5.712	5.739	5.766	5.794
270	5.821	5.848	5.875	5.903	5.930	5.957	5.985	6.012	6.040	6.067
280	6.094	6.122	6.149	6.177	6.204	6.232	6.259	6.287	6.314	6.342
290	6.370	6.397	6.425	6.453	6.481	6.508	6.536	6.564	6.592	6.620
300	6.647	6.675	6.703	6.731	6.759	6.786	6.814	6.842	6.870	6.898
310	6.926	6.954	6.982	7.010	7.038	7.066	7.095	7.123	7.151	7.180
320	7.208	7.236	7.264	7.292	7.321	7.349	7.377	7.405	7.434	7.462
330	7.491	7.519	7.548	7.576	7.605	7.633	7.661	7.690	7.719	7.747
340	7.776	7.805	7.834	7.862	7.891	7.920	7.949	7.978	8.006	8.035
350	8.064	8.092	8.120	8.149	8.178	8.207	8.236	8.265	8.294	8.323
360	8.352	8.381	8.410	8.439	8.468	8.497	8.526	8.555	8.584	8.613
370	8.642	8.672	8.701	8.730	8.759	8.788	8.818	8.847	8.876	8.905
380	8.935	8.964	8.994	9.023	9.052	9.082	9.111	9.141	9.170	9.200
390	9.229	9.259	9.288	9.317	9.347	9.376	9.406	9.436	9.466	9.495
400	9.525	9.555	9.584	9.614	9.644	9.674	9.703	9.733	9.763	9.793
410	9.823	9.853	9.883	9.913	9.943	9.973	10.003	10.033	10.063	10.093
420	10.123	10.153	10.183	10.213	10.243	10.273	10.303	10.333	10.363	10.393
430	10.423	10.453	10.483	10.514	10.544	10.574	10.604	10.635	10.665	10.695
440	10.726	10.756	10.787	10.817	10.848	10.878	10.909	10.939	10.969	11.000
450	11.030	11.061	11.091	11.122	11.152	11.183	11.214	11.244	11.275	11.305
460	11.336	11.366	11.397	11.428	11.459	11.490	11.520	11.551	11.581	11.612
470	11.643	11.674	11.704	11.735	11.766	11.797	11.828	11.859	11.891	11.922
480	11.953	11.984	12.015	12.046	12.077	12.108	12.138	12.170	12.201	12.232
490	12.263	12.294	12.325	12.356	12.387	12.418	12.450	12.481	12.512	12.543
500	12.575	12.606	12.637	12.669	12.700	12.732	12.763	12.794	12.825	12.857
510	12.888	12.919	12.951	12.983	13.014	13.046	13.077	13.108	13.140	13.172
520	13.203	13.235	13.267	13.298	13.330	13.362	13.393	13.425	13.457	13.488
530	13.520	13.552	13.583	13.615	13.647	13.678	13.710	13.742	13.774	13.806
540	13.838	13.869	13.901	13.933	13.965	13.997	14.029	14.061	14.093	14.125
550	14.157	14.189	14.221	14.253	14.285	14.317	14.349	14.381	14.413	14.445
560	14.477	14.509	14.541	14.573	14.605	14.637	14.670	14.702	14.734	14.766
570	14.799	14.831	14.864	14.896	14.928	14.961	14.993	15.025	15.057	15.090
580	15.122	15.155	15.187	15.219	15.252	15.284	15.317	15.349	15.382	15.414
590	15.447	15.480	15.512	15.545	15.577	15.610	15.642	15.675	15.707	15.740
600	15.773	15.806	15.838	15.871	15.904	15.937	15.969	16.002	16.035	16.068
610	16.101	16.133	16.166	16.199	16.232	16.264	16.297	16.330	16.363	16.396
620	16.429	16.462	16.495	16.528	16.560	16.593	16.626	16.659	16.692	16.725
630	16.758	16.791	16.824	16.857	16.890	16.924	16.957	16.990	17.023	17.056
640	17.089	17.122	17.155	17.189	17.222	17.255	17.288	17.321	17.354	17.388
650	17.421	17.454	17.488	17.521	17.554	17.588	17.621	17.654	17.688	17.721
660	17.754	17.788	17.821	17.854	17.888	17.921	17.955	17.988	18.022	18.055
670	18.089	18.123	18.156	18.190	18.223	18.257	18.290	18.324	18.357	18.391
680	18.425	18.458	18.492	18.526	18.560	18.593	18.627	18.660	18.694	18.727
690	18.761	18.795	18.829	18.863	18.896	18.930	18.964	18.998	19.032	19.066
700	19.100	19.134	19.168	19.201	19.235	19.269	19.303	19.337	19.371	19.405
710	19.439	19.473	19.506	19.540	19.574	19.608	19.642	19.676	19.711	19.745
720	19.779	19.813	19.847	19.881	19.915	19.949	19.983	20.018	20.052	20.086
730	20.120	20.154	20.188	20.223	20.257	20.291	20.325	20.359	20.394	20.428
740	20.463	20.497	20.531	20.565	20.599	20.634	20.668	20.702	20.736	20.771
750	20.805	20.840	20.874	....	....	....	....	....	....	....

Source: National Bureau of Standards Circular 561, 1955. (Reprinted by permission of Leeds & Northrup Company.)

TABLE A.5. **Output of type K thermocouple**

CHROMEL vs. ALUMEL THERMOCOUPLE
Degrees Fahrenheit   Reference Junction 32° F.

°F	0	1	2	3	4	5	6	7	8	9
					Millivolts					
0	-0.68	-0.66	-0.64	-0.62	-0.60	-0.58	-0.56	-0.54	-0.52	-0.49
10	-0.47	-0.45	-0.43	-0.41	-0.39	-0.37	-0.34	-0.32	-0.30	-0.28
20	-0.26	-0.24	-0.22	-0.19	-0.17	-0.15	-0.13	-0.11	-0.09	-0.07
30	-0.04	-0.02	0.00	0.02	0.04	0.07	0.09	0.11	0.13	0.15
40	0.18	0.20	0.22	0.24	0.26	0.29	0.31	0.33	0.35	0.37
50	0.40	0.42	0.44	0.46	0.48	0.51	0.53	0.55	0.57	0.60
60	0.62	0.64	0.66	0.68	0.71	0.73	0.75	0.77	0.80	0.82
70	0.84	0.86	0.88	0.91	0.93	0.95	0.97	1.00	1.02	1.04
80	1.06	1.09	1.11	1.13	1.15	1.18	1.20	1.22	1.24	1.27
90	1.29	1.31	1.33	1.36	1.38	1.40	1.43	1.45	1.47	1.49
100	1.52	1.54	1.56	1.58	1.61	1.63	1.65	1.68	1.70	1.72
110	1.74	1.77	1.79	1.81	1.84	1.86	1.88	1.90	1.93	1.95
120	1.97	2.00	2.02	2.04	2.06	2.09	2.11	2.13	2.16	2.18
130	2.20	2.23	2.25	2.27	2.29	2.32	2.34	2.36	2.39	2.41
140	2.43	2.46	2.48	2.50	2.52	2.55	2.57	2.59	2.62	2.64
150	2.66	2.69	2.71	2.73	2.75	2.78	2.80	2.82	2.85	2.87
160	2.89	2.92	2.94	2.96	2.98	3.01	3.03	3.05	3.08	3.10
170	3.12	3.15	3.17	3.19	3.22	3.24	3.26	3.29	3.31	3.33
180	3.36	3.38	3.40	3.43	3.45	3.47	3.49	3.52	3.54	3.56
190	3.59	3.61	3.63	3.66	3.68	3.70	3.73	3.75	3.77	3.80
200	3.82	3.84	3.87	3.89	3.91	3.94	3.96	3.98	4.01	4.03
210	4.05	4.08	4.10	4.12	4.15	4.17	4.19	4.21	4.24	4.26
220	4.28	4.31	4.33	4.35	4.38	4.40	4.42	4.44	4.47	4.49
230	4.51	4.54	4.56	4.58	4.61	4.63	4.65	4.67	4.70	4.72
240	4.74	4.77	4.79	4.81	4.83	4.86	4.88	4.90	4.92	4.95
250	4.97	4.99	5.02	5.04	5.06	5.08	5.11	5.13	5.15	5.17
260	5.20	5.22	5.24	5.26	5.29	5.31	5.33	5.35	5.38	5.40
270	5.42	5.44	5.47	5.49	5.51	5.53	5.56	5.58	5.60	5.62
280	5.65	5.67	5.69	5.71	5.73	5.76	5.78	5.80	5.82	5.85
290	5.87	5.89	5.91	5.93	5.96	5.98	6.00	6.02	6.05	6.07
300	6.09	6.11	6.13	6.16	6.18	6.20	6.22	6.25	6.27	6.29
310	6.31	6.33	6.36	6.38	6.40	6.42	6.45	6.47	6.49	6.51
320	6.53	6.56	6.58	6.60	6.62	6.65	6.67	6.69	6.71	6.73
330	6.76	6.78	6.80	6.82	6.84	6.87	6.89	6.91	6.93	6.96
340	6.98	7.00	7.02	7.04	7.07	7.09	7.11	7.13	7.15	7.18
350	7.20	7.22	7.24	7.26	7.29	7.31	7.33	7.35	7.38	7.40
360	7.42	7.44	7.46	7.49	7.51	7.53	7.55	7.58	7.60	7.62
370	7.64	7.66	7.69	7.71	7.73	7.75	7.78	7.80	7.82	7.84
380	7.87	7.89	7.91	7.93	7.95	7.98	8.00	8.02	8.04	8.07
390	8.09	8.11	8.13	8.16	8.18	8.20	8.22	8.24	8.27	8.29
400	8.31	8.33	8.36	8.38	8.40	8.42	8.45	8.47	8.49	8.51
410	8.54	8.56	8.58	8.60	8.62	8.65	8.67	8.69	8.71	8.74
420	8.76	8.78	8.80	8.82	8.85	8.87	8.89	8.91	8.94	8.96
430	8.98	9.00	9.03	9.05	9.07	9.09	9.12	9.14	9.16	9.18
440	9.21	9.23	9.25	9.27	9.30	9.32	9.34	9.36	9.39	9.41
450	9.43	9.45	9.48	9.50	9.52	9.54	9.57	9.59	9.61	9.63
460	9.66	9.68	9.70	9.73	9.75	9.77	9.79	9.82	9.84	9.86
470	9.88	9.91	9.93	9.95	9.97	10.00	10.02	10.04	10.06	10.09
480	10.11	10.13	10.16	10.18	10.20	10.22	10.25	10.27	10.29	10.31
490	10.34	10.36	10.38	10.40	10.43	10.45	10.47	10.50	10.52	10.54
500	10.57	10.59	10.61	10.63	10.66	10.68	10.70	10.72	10.75	10.77
510	10.79	10.82	10.84	10.86	10.88	10.91	10.93	10.95	10.98	11.00
520	11.02	11.04	11.07	11.09	11.11	11.13	11.16	11.18	11.20	11.23
530	11.25	11.27	11.29	11.32	11.34	11.36	11.39	11.41	11.43	11.45
540	11.48	11.50	11.52	11.55	11.57	11.59	11.61	11.64	11.66	11.68
550	11.71	11.73	11.75	11.78	11.80	11.82	11.84	11.87	11.89	11.91
560	11.94	11.96	11.98	12.01	12.03	12.05	12.07	12.10	12.12	12.14
570	12.17	12.19	12.21	12.24	12.26	12.28	12.30	12.33	12.35	12.37
580	12.40	12.42	12.44	12.47	12.49	12.51	12.53	12.56	12.58	12.60
590	12.63	12.65	12.67	12.70	12.72	12.74	12.76	12.79	12.81	12.83

TABLE A.5. (Continued)

CHROMEL vs. ALUMEL THERMOCOUPLE
Degrees Fahrenheit   Reference Junction 32° F.

°F	0	1	2	3	4	5	6	7	8	9
					Millivolts					
600	12.86	12.88	12.90	12.93	12.95	12.97	13.00	13.02	13.04	13.06
610	13.09	13.11	13.13	13.16	13.18	13.20	13.23	13.25	13.27	13.30
620	13.32	13.34	13.36	13.39	13.41	13.44	13.46	13.48	13.50	13.53
630	13.55	13.57	13.60	13.62	13.64	13.67	13.69	13.71	13.74	13.76
640	13.78	13.81	13.83	13.85	13.88	13.90	13.92	13.95	13.97	13.99
650	14.02	14.04	14.06	14.09	14.11	14.13	14.15	14.18	14.20	14.22
660	14.25	14.27	14.29	14.32	14.34	14.36	14.39	14.41	14.43	14.46
670	14.48	14.50	14.53	14.55	14.57	14.60	14.62	14.64	14.67	14.69
680	14.71	14.74	14.76	14.78	14.81	14.83	14.85	14.88	14.90	14.92
690	14.95	14.97	14.99	15.02	15.04	15.06	15.09	15.11	15.13	15.16
700	15.18	15.20	15.23	15.25	15.27	15.30	15.32	15.34	15.37	15.39
710	15.41	15.44	15.46	15.48	15.51	15.53	15.55	15.58	15.60	15.62
720	15.65	15.67	15.69	15.72	15.74	15.76	15.79	15.81	15.83	15.86
730	15.88	15.90	15.93	15.95	15.98	16.00	16.02	16.05	16.07	16.09
740	16.12	16.14	16.16	16.19	16.21	16.23	16.26	16.28	16.30	16.33
750	16.35	16.37	16.40	16.42	16.45	16.47	16.49	16.52	16.54	16.56
760	16.59	16.61	16.63	16.66	16.68	16.70	16.73	16.75	16.77	16.80
770	16.82	16.84	16.87	16.89	16.92	16.94	16.96	16.99	17.01	17.03
780	17.06	17.08	17.10	17.13	17.15	17.17	17.20	17.22	17.24	17.27
790	17.29	17.31	17.34	17.36	17.39	17.41	17.43	17.46	17.48	17.50
800	17.53	17.55	17.57	17.60	17.62	17.64	17.67	17.69	17.71	17.74
810	17.76	17.78	17.81	17.83	17.86	17.88	17.90	17.93	17.95	17.97
820	18.00	18.02	18.04	18.07	18.09	18.11	18.14	18.16	18.18	18.21
830	18.23	18.25	18.28	18.30	18.33	18.35	18.37	18.40	18.42	18.44
840	18.47	18.49	18.51	18.54	18.56	18.58	18.61	18.63	18.65	18.68
850	18.70	18.73	18.75	18.77	18.80	18.82	18.84	18.87	18.89	18.91
860	18.94	18.96	18.98	19.01	19.03	19.06	19.08	19.10	19.13	19.15
870	19.18	19.20	19.22	19.25	19.27	19.29	19.32	19.34	19.36	19.39
880	19.41	19.44	19.46	19.48	19.51	19.53	19.55	19.58	19.60	19.63
890	19.65	19.67	19.70	19.72	19.75	19.77	19.79	19.82	19.84	19.86
900	19.89	19.91	19.94	19.96	19.98	20.01	20.03	20.05	20.08	20.10
910	20.13	20.15	20.17	20.20	20.22	20.24	20.27	20.29	20.32	20.34
920	20.36	20.39	20.41	20.43	20.46	20.48	20.50	20.53	20.55	20.58
930	20.60	20.62	20.65	20.67	20.69	20.72	20.74	20.76	20.79	20.81
940	20.84	20.86	20.88	20.91	20.93	20.95	20.98	21.00	21.03	21.05
950	21.07	21.10	21.12	21.14	21.17	21.19	21.21	21.24	21.26	21.28
960	21.31	21.33	21.36	21.38	21.40	21.43	21.45	21.47	21.50	21.52
970	21.54	21.57	21.59	21.62	21.64	21.66	21.69	21.71	21.73	21.76
980	21.78	21.81	21.83	21.85	21.88	21.90	21.92	21.95	21.97	21.99
990	22.02	22.04	22.07	22.09	22.11	22.14	22.16	22.18	22.21	22.23
1000	22.26	22.28	22.30	22.33	22.35	22.37	22.40	22.42	22.44	22.47
1010	22.49	22.52	22.54	22.56	22.59	22.61	22.63	22.66	22.68	22.71
1020	22.73	22.75	22.78	22.80	22.82	22.85	22.87	22.90	22.92	22.94
1030	22.97	22.99	23.01	23.04	23.06	23.08	23.11	23.13	23.16	23.18
1040	23.20	23.23	23.25	23.27	23.30	23.32	23.35	23.37	23.39	23.42
1050	23.44	23.46	23.49	23.51	23.54	23.56	23.58	23.61	23.63	23.65
1060	23.68	23.70	23.72	23.75	23.77	23.80	23.82	23.84	23.87	23.89
1070	23.91	23.94	23.96	23.99	24.01	24.03	24.06	24.08	24.10	24.13
1080	24.15	24.18	24.20	24.22	24.25	24.27	24.29	24.32	24.34	24.36
1090	24.39	24.41	24.44	24.46	24.49	24.51	24.53	24.55	24.58	24.60
1100	24.63	24.65	24.67	24.70	24.72	24.74	24.77	24.79	24.82	24.84
1110	24.86	24.89	24.91	24.93	24.96	24.98	25.01	25.03	25.05	25.08
1120	25.10	25.12	25.15	25.17	25.20	25.22	25.24	25.27	25.29	25.31
1130	25.34	25.36	25.38	25.41	25.43	25.46	25.48	25.50	25.53	25.55
1140	25.57	25.60	25.62	25.65	25.67	25.69	25.72	25.74	25.76	25.79
1150	25.81	25.83	25.86	25.88	25.91	25.93	25.95	25.98	26.00	26.02
1160	26.05	26.07	26.09	26.12	26.14	26.16	26.19	26.21	26.24	26.26
1170	26.28	26.31	26.33	26.35	26.38	26.40	26.42	26.45	26.47	26.49
1180	26.52	26.54	26.56	26.59	26.61	26.63	26.66	26.68	26.70	26.73
1190	26.75	26.77	26.80	26.82	26.85	26.87	26.89	26.91	26.94	26.96

TABLE A.5. (Continued)

CHROMEL vs. ALUMEL THERMOCOUPLE
Degrees Fahrenheit   Reference Junction 32° F.

°F	0	1	2	3	4	5	6	7	8	9
					Millivolts					
1200	26.98	27.01	27.03	27.06	27.08	27.10	27.12	27.15	27.17	27.20
1210	27.22	27.24	27.27	27.29	27.31	27.34	27.36	27.38	27.40	27.43
1220	27.45	27.48	27.50	27.52	27.55	27.57	27.59	27.62	27.64	27.66
1230	27.69	27.71	27.73	27.76	27.78	27.80	27.83	27.85	27.87	27.90
1240	27.92	27.94	27.97	27.99	28.01	28.04	28.06	28.08	28.11	28.13
1250	28.15	28.18	28.20	28.22	28.25	28.27	28.29	28.32	28.34	28.37
1260	28.39	28.41	28.44	28.46	28.48	28.50	28.53	28.55	28.58	28.60
1270	28.62	28.65	28.67	28.69	28.72	28.74	28.76	28.79	28.81	28.83
1280	28.86	28.88	28.90	28.93	28.95	28.97	29.00	29.02	29.04	29.07
1290	29.09	29.11	29.14	29.16	29.18	29.21	29.23	29.25	29.28	29.30
1300	29.32	29.35	29.37	29.39	29.42	29.44	29.46	29.49	29.51	29.53
1310	29.56	29.58	29.60	29.63	29.65	29.67	29.70	29.72	29.74	29.77
1320	29.79	29.81	29.84	29.86	29.88	29.91	29.93	29.95	29.97	30.00
1330	30.02	30.05	30.07	30.09	30.11	30.14	30.16	30.18	30.21	30.23
1340	30.25	30.28	30.30	30.32	30.35	30.37	30.39	30.42	30.44	30.46
1350	30.49	30.51	30.53	30.56	30.58	30.60	30.63	30.65	30.67	30.70
1360	30.72	30.74	30.77	30.79	30.81	30.83	30.86	30.88	30.90	80.93
1370	30.95	30.97	31.00	31.02	31.04	31.07	31.09	31.11	31.14	31.16
1380	31.18	31.21	31.23	31.25	31.28	31.30	31.32	31.34	31.37	31.39
1390	31.42	31.44	31.46	31.48	31.51	31.53	31.55	31.58	31.60	31.62
1400	31.65	31.67	31.69	31.72	31.74	31.76	31.78	31.81	31.83	31.85
1410	31.88	31.90	31.92	31.95	31.97	31.99	32.02	32.04	32.06	32.08
1420	32.11	32.13	32.15	32.18	32.20	32.22	32.25	32.27	32.29	32.31
1430	32.34	32.36	32.38	32.41	32.43	32.45	32.48	32.50	32.52	32.54
1440	32.57	32.59	32.61	32.64	32.66	32.68	32.70	32.73	32.75	32.77
1450	32.80	32.82	32.84	32.86	32.89	32.91	32.93	32.96	32.98	33.00
1460	33.02	33.05	33.07	33.09	33.12	33.14	33.16	33.18	33.21	33.23
1470	33.25	33.28	33.30	33.32	33.34	33.37	33.39	33.41	33.43	33.46
1480	33.48	33.50	33.53	33.55	33.57	33.59	33.62	33.64	33.66	33.69
1490	33.71	33.73	33.75	33.78	33.80	33.82	33.84	33.87	33.89	33.91
1500	33.93	33.96	33.98	34.00	34.03	34.05	34.07	34.09	34.12	34.14
1510	34.16	34.18	34.21	34.23	34.25	34.28	34.30	34.32	34.34	34.37
1520	34.39	34.41	34.43	34.46	34.48	34.50	34.53	34.55	34.57	34.59
1530	34.62	34.64	34.66	34.68	34.71	34.73	34.75	34.77	34.80	34.82
1540	34.84	34.87	34.89	34.91	34.93	34.96	34.98	35.00	35.02	35.05
1550	35.07	35.09	35.11	35.14	35.16	35.18	35.21	35.23	35.25	35.27
1560	35.29	35.32	35.34	35.36	35.39	35.41	35.43	35.45	35.48	35.50
1570	35.52	35.54	35.57	35.59	35.61	35.63	35.66	35.68	35.70	35.72
1580	35.75	35.77	35.79	35.81	35.84	35.86	35.88	35.90	35.93	35.95
1590	35.97	35.99	36.02	36.04	36.06	36.08	36.11	36.13	36.15	36.17
1600	36.19	36.22	36.24	36.26	36.29	36.31	36.33	36.35	36.37	36.40
1610	36.42	36.44	36.46	36.49	36.51	36.53	36.55	36.58	36.60	36.62
1620	36.64	36.67	36.69	36.71	36.73	36.76	36.78	36.80	36.82	36.84
1630	36.87	36.89	36.91	36.93	36.96	36.98	37.00	37.02	37.05	37.07
1640	37.09	37.11	37.14	37.16	37.18	37.20	37.23	37.25	37.27	37.29
1650	37.31	37.34	37.36	37.38	37.40	37.43	37.45	37.47	37.49	37.52
1660	37.54	37.56	37.58	37.60	37.63	37.65	37.67	37.69	37.72	37.74
1670	37.76	37.78	37.81	37.83	37.85	37.87	37.89	37.92	37.94	37.96
1680	37.98	38.01	38.03	38.05	38.07	38.09	38.12	38.14	38.16	38.18
1690	38.20	38.23	38.25	38.27	38.29	38.32	38.34	38.36	38.38	38.40
1700	38.43	38.45	38.47	38.49	38.51	38.54	38.56	38.58	38.60	38.62
1710	38.65	38.67	38.69	38.71	38.73	38.76	38.78	38.80	38.82	38.84
1720	38.87	38.89	38.91	38.93	38.95	38.98	39.00	39.02	39.04	39.06
1730	39.09	39.11	39.13	39.15	39.17	39.20	39.22	39.24	39.26	39.28
1740	39.31	39.33	39.35	39.37	39.39	39.42	39.44	39.46	39.48	39.50
1750	39.53	39.55	39.57	39.59	39.61	39.64	39.66	39.68	39.70	39.72
1760	39.75	39.77	39.79	39.81	39.83	39.86	39.88	39.90	39.92	39.94
1770	39.96	39.99	40.01	40.03	40.05	40.07	40.10	40.12	40.14	40.16
1780	40.18	40.20	40.23	40.25	40.27	40.29	40.31	40.34	40.36	40.38
1790	40.40	40.42	40.44	40.47	40.49	40.51	40.53	40.55	40.58	40.60

TABLE A.5. **(Continued)**

CHROMEL vs. ALUMEL THERMOCOUPLE
Degrees Fahrenheit   Reference Junction 32° F.

°F	0	1	2	3	4	5	6	7	8	9
					Millivolts					
1800	40.62	40.64	40.66	40.68	40.71	40.73	40.75	40.77	40.79	40.82
1810	40.84	40.86	40.88	40.90	40.92	40.95	40.97	40.99	41.01	41.03
1820	41.05	41.08	41.10	41.12	41.14	41.16	41.18	41.21	41.23	41.25
1830	41.27	41.29	41.31	41.34	41.36	41.38	41.40	41.42	41.45	41.47
1840	41.49	41.51	41.53	41.55	41.57	41.60	41.62	41.64	41.66	41.68
1850	41.70	41.73	41.75	41.77	41.79	41.81	41.83	41.85	41.88	41.90
1860	41.92	41.94	41.96	41.99	42.01	42.03	42.05	42.07	42.09	42.11
1870	42.14	42.16	42.18	42.20	42.22	42.24	42.26	42.29	42.31	42.33
1880	42.35	42.37	42.39	42.42	42.44	42.46	42.48	42.50	42.52	42.55
1890	42.57	42.59	42.61	42.63	42.65	42.67	42.69	42.72	42.74	42.76
1900	42.78	42.80	42.82	42.84	42.87	42.89	42.91	42.93	42.95	42.97
1910	42.99	43.01	43.04	43.06	43.08	43.10	43.12	43.14	43.17	43.19
1920	43.21	43.23	43.25	43.27	43.29	43.31	43.34	43.36	43.38	43.40
1930	43.42	43.44	43.47	43.49	43.51	43.53	43.55	43.57	43.59	43.61
1940	43.63	43.66	43.68	43.70	43.72	43.74	43.76	43.78	43.81	43.83
1950	43.85	43.87	43.89	43.91	43.93	43.95	43.98	44.00	44.02	44.04
1960	44.06	44.08	44.10	44.13	44.15	44.17	44.19	44.21	44.23	44.25
1970	44.27	44.30	44.32	44.34	44.36	44.38	44.40	44.42	44.44	44.47
1980	44.49	44.51	44.53	44.55	44.57	44.59	44.61	44.63	44.66	44.68
1990	44.70	44.72	44.74	44.76	44.78	44.80	44.82	44.85	44.87	44.89
2000	44.91	44.93	44.95	44.97	44.99	45.01	45.03	45.06	45.08	45.10
2010	45.12	45.14	45.16	45.18	45.20	45.22	45.24	45.27	45.29	45.31
2020	45.33	45.35	45.37	45.39	45.41	45.43	45.45	45.48	45.50	45.52
2030	45.54	45.56	45.58	45.60	45.62	45.64	45.66	45.69	45.71	45.73
2040	45.75	45.77	45.79	45.81	45.83	45.85	45.87	45.90	45.92	45.94
2050	45.96	45.98	46.00	46.02	46.04	46.06	46.08	46.11	46.13	46.15
2060	46.17	46.19	46.21	46.23	46.25	46.27	46.29	46.31	46.33	46.36
2070	46.38	46.40	46.42	46.44	46.46	46.48	46.50	46.52	46.54	46.56
2080	46.58	46.60	46.63	46.65	46.67	46.69	46.71	46.73	46.75	46.77
2090	46.79	46.81	46.83	46.85	46.87	46.90	46.92	46.94	46.96	46.98
2100	47.00	47.02	47.04	47.06	47.08	47.10	47.12	47.14	47.17	47.19
2110	47.21	47.23	47.25	47.27	47.29	47.31	47.33	47.35	47.37	47.39
2120	47.41	47.43	47.45	47.47	47.49	47.52	47.54	47.56	47.58	47.60
2130	47.62	47.64	47.66	47.68	47.70	47.72	47.74	47.76	47.78	47.80
2140	47.82	47.84	47.86	47.89	47.91	47.93	47.95	47.97	47.99	48.01
2150	48.03	48.05	48.07	48.09	48.11	48.13	48.15	48.17	48.19	48.21
2160	48.23	48.25	48.27	48.29	48.32	48.34	48.36	48.38	48.40	48.42
2170	48.44	48.46	48.48	48.50	48.52	48.54	48.56	48.58	48.60	48.62
2180	48.64	48.66	48.68	48.70	48.72	48.74	48.76	48.79	48.81	48.83
2190	48.85	48.87	48.89	48.91	48.93	48.95	48.97	48.99	49.01	49.03
2200	49.05	49.07	49.09	49.11	49.13	49.15	49.17	49.19	49.21	49.23
2210	49.25	49.27	49.29	49.31	49.33	49.35	49.37	49.39	49.41	49.43
2220	49.45	49.47	49.49	49.51	49.53	49.56	49.57	49.59	49.61	49.63
2230	49.65	49.67	49.69	49.71	49.73	49.76	49.78	49.80	49.82	49.84
2240	49.86	49.88	49.90	49.92	49.94	49.96	49.98	50.00	50.02	50.04
2250	50.06	50.08	50.10	50.12	50.14	50.16	50.18	50.20	50.22	50.24
2260	50.26	50.28	50.30	50.32	50.34	50.36	50.38	50.40	50.42	50.44
2270	50.46	50.48	50.50	50.52	50.54	50.56	50.57	50.59	50.61	50.63
2280	50.65	50.67	50.69	50.71	50.73	50.75	50.77	50.79	50.81	50.83
2290	50.85	50.87	50.89	50.91	50.93	50.95	50.97	50.99	51.01	51.03
2300	51.05	51.07	51.09	51.11	51.13	51.15	51.17	51.19	51.21	51.23
2310	51.25	51.27	51.29	51.31	51.33	51.35	51.37	51.39	51.41	51.43
2320	51.45	51.47	51.48	51.50	51.52	51.54	51.56	51.58	51.60	51.62
2330	51.64	51.66	51.68	51.70	51.72	51.74	51.76	51.78	51.80	51.82
2340	51.84	51.86	51.88	51.90	51.92	51.94	51.96	51.98	52.00	51.01
2350	52.03	52.05	52.07	52.09	52.11	52.13	52.15	52.17	52.19	52.21
2360	52.23	52.25	52.27	52.29	52.31	52.33	52.35	52.37	52.39	52.41
2370	52.42	52.44	52.46	52.48	52.50	52.52	52.54	52.56	52.58	52.60
2380	52.62	52.64	52.66	52.68	52.70	52.72	52.74	52.76	52.77	52.79
2390	52.81	52.83	52.85	52.87	52.89	52.91	52.93	52.95	52.97	52.99

TABLE A.5. (Continued)

CHROMEL vs. ALUMEL THERMOCOUPLE
Degrees Fahrenheit   Reference Junction 32° F.

°F	0	1	2	3	4	5	6	7	8	9
					Millivolts					
2400	53.01	53.03	53.05	53.07	53.08	53.10	53.12	53.14	53.16	53.18
2410	53.20	53.22	53.24	53.26	53.28	53.30	53.32	53.34	53.35	53.37
2420	53.39	53.41	53.43	53.45	53.47	53.49	53.51	53.53	53.55	53.57
2430	53.59	53.60	53.62	53.64	53.66	53.68	53.70	53.72	53.74	53.76
2440	53.78	53.80	53.82	53.83	53.85	53.87	53.89	53.91	53.93	53.95
2450	53.97	53.99	54.01	54.03	54.04	54.06	54.08	54.10	54.12	54.14
2460	54.16	54.18	54.20	54.22	54.24	54.25	54.27	54.29	54.31	54.33
2470	54.35	54.37	54.39	54.41	54.43	54.44	54.46	54.48	54.50	54.52
2480	54.54	54.56	54.58	54.60	54.62	54.63	54.65	54.67	54.69	54.71
2490	54.73	54.75	54.77	54.79	54.81	54.82	54.84	54.86	54.88	54.90

Source: National Bureau of Standards Circular 561, 1955. (Reprinted by permission of Leeds & Northrup Company.)

TABLE A.6. Output of type S thermocouple

PLAT. +10% RHODIUM vs. PLAT. THERMOCOUPLE
Degrees Fahrenheit   Reference Junction 32° F.

°F	0	1	2	3	4	5	6	7	8	9
					Millivolts					
0	-0.093	-0.090	-0.087	-0.085	-0.082	-0.079	-0.076	-0.073	-0.071	-0.068
10	-.065	-.062	-.059	-.056	-.053	-.051	-.048	-.045	-.042	-.039
20	-.036	-.033	-.030	-.027	-.024	-.021	-.018	-.015	-.012	-.009
30	-.006	-.003	.000	.003	.006	.009	.012	.015	.018	.021
40	.024	.028	.031	.034	.037	.040	.043	.046	.049	.052
50	.056	.059	.062	.065	.068	.071	.075	.078	.081	.084
60	.087	.091	.094	.097	.100	.104	.107	.110	.113	.117
70	.120	.123	.126	.130	.133	.136	.140	.143	.146	.150
80	.153	.156	.160	.163	.166	.170	.173	.176	.180	.183
90	.187	.190	.193	.197	.200	.204	.207	.211	.214	.218
100	.221	.224	.228	.231	.235	.238	.242	.245	.249	.252
110	.256	.259	.263	.266	.270	.274	.277	.281	.284	.288
120	.291	.295	.299	.302	.306	.309	.313	.317	.320	.324
130	.327	.331	.335	.338	.342	.346	.349	.353	.357	.360
140	.364	.368	.371	.375	.379	.383	.386	.390	.394	.397
150	.401	.405	.409	.412	.416	.420	.424	.428	.431	.435
160	.439	.443	.447	.450	.454	.458	.462	.466	.469	.473
170	.477	.481	.485	.489	.493	.496	.500	.504	.508	.512
180	.516	.520	.524	.528	.532	.535	.539	.543	.547	.551
190	.555	.559	.563	.567	.571	.575	.579	.583	.587	.591
200	.595	.599	.603	.607	.611	.615	.619	.623	.627	.631
210	.635	.639	.643	.647	.651	.655	.659	.664	.668	.672
220	.676	.680	.684	.688	.692	.696	.700	.705	.709	.713
230	.717	.721	.725	.729	.734	.738	.742	.746	.750	.754
240	.758	.763	.767	.771	.775	.779	.784	.788	.792	.796
250	.800	.805	.809	.813	.817	.822	.826	.830	.834	.839
260	.843	.847	.851	.856	.860	.864	.869	.873	.877	.881
270	.886	.890	.894	.899	.903	.907	.912	.916	.920	.925
280	.929	.933	.938	.942	.946	.951	.955	.959	.964	.968
290	.973	.977	.981	.986	.990	.994	.999	1.003	1.008	1.012
300	1.017	1.021	1.025	1.030	1.034	1.039	1.043	1.048	1.052	1.056
310	1.061	1.065	1.070	1.074	1.079	1.083	1.088	1.092	1.097	1.101
320	1.106	1.110	1.115	1.119	1.124	1.128	1.132	1.137	1.142	1.146
330	1.151	1.155	1.160	1.164	1.169	1.173	1.178	1.182	1.187	1.191
340	1.196	1.200	1.205	1.210	1.214	1.219	1.223	1.228	1.232	1.237
350	1.242	1.246	1.251	1.255	1.260	1.264	1.269	1.274	1.278	1.283
360	1.287	1.292	1.297	1.301	1.306	1.311	1.315	1.320	1.324	1.329
370	1.334	1.338	1.343	1.348	1.352	1.357	1.362	1.366	1.371	1.376
380	1.380	1.385	1.390	1.394	1.399	1.404	1.408	1.413	1.418	1.422
390	1.427	1.432	1.436	1.441	1.446	1.450	1.455	1.460	1.465	1.469
400	1.474	1.479	1.483	1.488	1.493	1.498	1.502	1.507	1.512	1.516
410	1.521	1.526	1.531	1.535	1.540	1.545	1.550	1.554	1.559	1.564
420	1.569	1.573	1.578	1.583	1.588	1.593	1.597	1.602	1.607	1.612
430	1.616	1.621	1.626	1.631	1.636	1.640	1.645	1.650	1.655	1.660
440	1.664	1.669	1.674	1.679	1.684	1.688	1.693	1.698	1.703	1.708
450	1.712	1.717	1.722	1.727	1.732	1.736	1.741	1.746	1.751	1.756
460	1.761	1.765	1.770	1.775	1.780	1.785	1.790	1.795	1.799	1.804
470	1.809	1.814	1.819	1.824	1.829	1.833	1.838	1.843	1.848	1.853
480	1.858	1.863	1.868	1.873	1.877	1.882	1.887	1.892	1.897	1.902
490	1.907	1.912	1.917	1.922	1.927	1.931	1.936	1.941	1.946	1.951
500	1.956	1.961	1.966	1.971	1.976	1.981	1.986	1.991	1.996	2.000
510	2.005	2.010	2.015	2.020	2.025	2.030	2.035	2.040	2.045	2.050
520	2.055	2.060	2.065	2.070	2.075	2.080	2.085	2.090	2.095	2.100
530	2.105	2.110	2.115	2.120	2.125	2.130	2.135	2.140	2.145	2.150
540	2.155	2.160	2.165	2.170	2.175	2.180	2.185	2.190	2.195	2.200
550	2.205	2.210	2.215	2.220	2.225	2.230	2.235	2.240	2.245	2.250
560	2.255	2.260	2.265	2.270	2.276	2.281	2.286	2.291	2.296	2.301
570	2.306	2.311	2.316	2.321	2.326	2.331	2.336	2.341	2.346	2.351
580	2.357	2.362	2.367	2.372	2.377	2.382	2.387	2.392	2.397	2.402
590	2.407	2.413	2.418	2.423	2.428	2.433	2.438	2.443	2.448	2.453

TABLE A.6. (Continued)

PLAT. +10% RHODIUM vs. PLAT. THERMOCOUPLE
Degrees Fahrenheit   Reference Junction 32° F.

°F	0	1	2	3	4	5	6	7	8	9
					Millivolts					
600	2.458	2.464	2.469	2.474	2.479	2.484	2.489	2.494	2.499	2.505
610	2.510	2.515	2.520	2.525	2.530	2.535	2.540	2.546	2.551	2.556
620	2.561	2.566	2.571	2.576	2.582	2.587	2.592	2.597	2.602	2.607
630	2.613	2.618	2.623	2.628	2.633	2.638	2.644	2.649	2.654	2.659
640	2.664	2.669	2.675	2.680	2.685	2.690	2.695	2.700	2.706	2.711
650	2.716	2.721	2.726	2.731	2.737	2.742	2.747	2.752	2.757	2.763
660	2.768	2.773	2.778	2.783	2.789	2.794	2.799	2.804	2.809	2.815
670	2.820	2.825	2.830	2.836	2.841	2.846	2.851	2.856	2.862	2.867
680	2.872	2.877	2.882	2.888	2.893	2.898	2.903	2.909	2.914	2.919
690	2.924	2.930	2.935	2.940	2.945	2.951	2.956	2.961	2.966	2.972
700	2.977	2.982	2.987	2.992	2.998	3.003	3.008	3.014	3.019	3.024
710	3.029	3.035	3.040	3.045	3.050	3.056	3.061	3.066	3.071	3.077
720	3.082	3.087	3.092	3.098	3.102	3.108	3.114	3.119	3.124	3.129
730	3.135	3.140	3.145	3.150	3.156	3.161	3.166	3.172	3.177	3.182
740	3.188	3.193	3.198	3.203	3.209	3.214	3.219	3.225	3.230	3.235
750	3.240	3.246	3.251	3.256	3.262	3.267	3.272	3.278	3.283	3.288
760	3.293	3.299	3.304	3.309	3.315	3.320	3.325	3.331	3.336	3.341
770	3.347	3.352	3.357	3.363	3.368	3.373	3.378	3.384	3.389	3.394
780	3.400	3.405	3.410	3.416	3.421	3.426	3.432	3.437	3.442	3.448
790	3.453	3.458	3.464	3.469	3.474	3.480	3.485	3.490	3.496	3.501
800	3.506	3.512	3.517	3.522	3.528	3.533	3.539	3.544	3.549	3.555
810	3.560	3.565	3.571	3.576	3.581	3.587	3.592	3.597	3.603	3.608
820	3.614	3.619	3.624	3.630	3.635	3.640	3.646	3.651	3.656	3.662
830	3.667	3.673	3.678	3.683	3.689	3.694	3.699	3.705	3.710	3.716
840	3.721	3.726	3.732	3.737	3.743	3.748	3.753	3.759	3.764	3.769
850	3.775	3.780	3.786	3.791	3.796	3.802	3.807	3.813	3.818	3.823
960	3.829	3.834	3.840	3.845	3.850	3.856	3.861	3.867	3.872	3.878
870	3.883	3.888	3.894	3.899	3.905	3.910	3.915	3.921	3.926	3.932
880	3.937	3.943	3.948	3.953	3.959	3.964	3.970	3.975	3.981	3.986
890	3.991	3.997	4.002	4.008	4.013	4.019	4.024	4.030	4.035	4.040
900	4.046	4.051	4.057	4.062	4.068	4.073	4.079	4.084	4.089	4.095
910	4.100	4.106	4.111	4.117	4.122	4.128	4.133	4.139	4.144	4.149
920	4.155	4.160	4.166	4.171	4.177	4.182	4.188	4.193	4.199	4.204
930	4.210	4.215	4.221	4.226	4.232	4.237	4.243	4.248	4.254	4.259
940	4.264	4.270	4.275	4.281	4.286	4.292	4.297	4.303	4.308	4.314
950	4.319	4.325	4.330	4.336	4.341	4.347	4.352	4.358	4.363	4.369
960	4.374	4.380	4.385	4.391	4.396	4.402	4.408	4.413	4.419	4.424
970	4.430	4.435	4.441	4.446	4.452	4.457	4.463	4.468	4.474	4.479
980	4.485	4.490	4.496	4.501	4.507	4.512	4.518	4.524	4.529	4.535
990	4.540	4.546	4.551	4.557	4.562	4.568	4.573	4.579	4.584	4.590
1000	4.596	4.601	4.607	4.612	4.618	4.623	4.629	4.634	4.640	4.646
1010	4.651	4.657	4.662	4.668	4.673	4.679	4.685	4.690	4.696	4.701
1020	4.707	4.712	4.718	4.724	4.729	4.735	4.740	4.746	4.751	4.757
1030	4.763	4.768	4.774	4.779	4.785	4.790	4.796	4.802	4.807	4.813
1040	4.818	4.824	4.830	4.835	4.841	4.846	4.852	4.858	4.863	4.869
1050	4.874	4.880	4.886	4.891	4.897	4.902	4.908	4.914	4.919	4.925
1060	4.930	4.936	4.942	4.947	4.953	4.959	4.964	4.970	4.975	4.981
1070	4.987	4.992	4.998	5.004	5.009	5.015	5.020	5.026	5.032	5.037
1080	5.043	5.049	5.054	5.060	5.066	5.071	5.077	5.082	5.088	5.094
1090	5.099	5.105	5.111	5.116	5.122	5.128	5.133	5.139	5.145	5.150
1100	5.156	5.162	5.167	5.173	5.178	5.184	5.190	5.195	5.201	5.207
1110	5.212	5.218	5.224	5.229	5.235	5.241	5.246	5.252	5.258	5.264
1120	5.269	5.275	5.281	5.286	5.292	5.298	5.303	5.309	5.315	5.320
1130	5.326	5.332	5.337	5.343	5.349	5.354	5.360	5.366	5.372	5.377
1140	5.383	5.389	5.394	5.400	5.406	5.411	5.417	5.423	5.429	5.434
1150	5.440	5.446	5.451	5.457	5.463	5.469	5.474	5.480	5.486	5.491
1160	5.497	5.503	5.509	5.514	5.520	5.526	5.532	5.537	5.543	5.549
1170	5.555	5.560	5.566	5.572	5.577	5.583	5.589	5.595	5.600	5.606
1180	5.612	5.617	5.623	5.629	5.635	5.640	5.646	5.652	5.658	5.663
1190	5.669	5.675	5.681	5.686	5.692	5.698	5.704	5.709	5.715	5.721

TABLE A.6. (Continued)

PLAT. +10% RHODIUM vs. PLAT. THERMOCOUPLE
Degrees Fahrenheit   Reference Junction 32° F.

°F	0	1	2	3	4	5	6	7	8	9
					Millivolts					
1200	5.726	5.732	5.738	5.744	5.749	5.755	5.761	5.767	5.773	5.778
1210	5.784	5.790	5.796	5.801	5.807	·5.813	5.819	5.824	5.830	5.836
1220	5.842	5.847	5.853	5.859	5.865	5.871	5.876	5.882	5.888	5.894
1230	5.899	5.905	5.911	5.917	5.923	5.928	5.934	5.940	5.946	5.951
1240	5.957	5.963	5.969	5.975	5.980	5.986	5.992	5.998	6.004	6.009
1250	6.015	6.021	6.027	6.033	6.038	6.044	6.050	6.056	6.062	6.067
1260	6.073	6.079	6.085	6.091	6.096	6.102	6.108	6.114	6.120	6.126
1270	6.131	6.137	6.143	6.149	6.155	6.161	6.166	6.172	6.178	6.184
1280	6.190	6.196	6.201	6.207	6.213	6.219	6.225	6.231	6.236	6.242
1290	6.248	6.254	6.260	6.266	6.271	6.277	6.283	6.289	6.295	6.301
1300	6.307	6.312	6.318	6.324	6.330	6.336	6.342	6.348	6.353	6.359
1310	6.365	6.371	6.377	6.383	6.389	6.394	6.400	6.406	6.412	6.418
1320	6.424	6.430	6.436	6.441	6.447	6.453	6.459	6.465	6.471	6.477
1330	6.483	6.488	6.494	6.500	6.506	6.512	6.518	6.524	6.530	6.536
1340	6.542	6.547	6.553	6.559	6.565	6.571	6.577	6.583	6.589	6.595
1350	6.601	6.606	6.612	6.618	6.624	6.630	6.636	6.642	6.648	6.654
1360	6.660	6.666	6.671	6.677	6.683	6.689	6.695	6.701	6.707	6.713
1370	6.719	6.725	6.731	6.737	6.743	6.749	6.754	6.760	6.766	6.772
1380	6.778	6.784	6.790	6.796	6.802	6.808	6.814	6.820	6.826	6.832
1390	6.838	6.844	6.850	6.855	6.861	6.867	6.873	6.879	6.885	6.891
1400	6.897	6.903	6.909	6.915	6.921	6.927	6.933	6.939	6.945	6.951
1410	6.957	6.963	6.969	6.975	6.981	6.987	6.993	6.999	7.005	7.011
1420	7.017	7.023	7.029	7.034	7.040	7.046	7.052	7.058	7.064	7.070
1430	7.076	7.082	7.088	7.094	7.100	7.106	7.112	7.118	7.124	7.130
1440	7.136	7.142	7.148	7.154	7.160	7.166	7.172	7.178	7.184	7.190
1450	7.196	7.202	7.208	7.214	7.220	7.226	7.233	7.239	7.245	7.251
1460	7.257	7.263	7.269	7.275	7.281	7.287	7.293	7.299	7.305	7.311
1470	7.317	7.323	7.329	7.335	7.341	7.347	7.353	7.359	7.365	7.371
1480	7.377	7.383	7.389	7.395	7.401	7.407	7.414	7.420	7.426	7.432
1490	7.438	7.444	7.450	7.456	7.462	7.468	7.474	7.480	7.486	7.492
1500	7.498	7.504	7.510	7.517	7.523	7.529	7.535	7.541	7.547	7.553
1510	7.559	7.565	7.571	7.577	7.583	7.589	7.596	7.602	7.608	7.614
1520	7.620	7.626	7.632	7.638	7.644	7.650	7.656	7.662	7.669	7.675
1530	7.681	7.687	7.693	7.699	7.705	7.711	7.717	7.723	7.730	7.736
1540	7.742	7.748	7.754	7.760	7.766	7.772	7.778	7.785	7.791	7.797
1550	7.803	7.809	7.815	7.821	7.827	7.834	7.840	7.846	7.852	7.858
1560	7.864	7.870	7.876	7.882	7.889	7.895	7.901	7.907	7.913	7.919
1570	7.925	7.932	7.938	7.944	7.950	7.956	7.962	7.968	7.975	7.981
1580	7.987	7.993	7.999	8.005	8.012	8.018	8.024	8.030	8.036	8.042
1590	8.048	8.055	8.061	8.067	8.073	8.079	8.085	8.092	8.098	8.104
1600	8.110	8.116	8.122	8.129	8.135	8.141	8.147	8.153	8.159	8.166
1610	8.172	8.178	8.184	8.190	8.197	8.203	8.209	8.215	8.221	8.228
1620	8.234	8.240	8.246	8.252	8.258	8.265	8.271	8.277	8.283	8.289
1630	8.296	8.302	8.308	8.314	8.320	8.327	8.333	8.339	8.345	8.352
1640	8.358	8.364	8.370	8.376	8.383	8.389	8.395	8.401	8.407	8.414
1650	8.420	8.426	8.432	8.439	8.445	8.451	8.457	8.464	8.470	8.476
1660	8.482	8.488	8.495	8.501	8.507	8.513	8.520	8.526	8.532	8.538
1670	8.545	8.551	8.557	8.563	8.570	8.576	8.582	8.588	8.595	8.601
1680	8.607	8.613	8.620	8.626	8.632	8.638	8.645	8.651	8.657	8.663
1690	8.670	8.676	8.682	8.689	8.695	8.701	8.707	8.714	8.720	8.726
1700	8.732	8.739	8.745	8.751	8.758	8.764	8.770	8.776	8.783	8.789
1710	8.795	8.802	8.808	8.814	8.820	8.827	8.833	8.839	8.846	8.852
1720	8.858	8.864	8.871	8.877	8.883	8.890	8.896	8.902	8.909	8.915
1730	8.921	8.927	8.934	8.940	8.946	8.953	8.959	8.965	8.972	8.978
1740	8.984	8.991	8.997	9.003	9.010	9.016	9.022	9.029	9.035	9.041
1750	9.048	9.054	9.060	9.067	9.073	9.079	9.086	9.092	9.098	9.105
1760	9.111	9.117	9.124	9.130	9.136	9.143	9.149	9.155	9.162	9.168
1770	9.174	9.181	9.187	9.193	9.200	9.206	9.212	9.219	9.225	9.232
1780	9.238	9.244	9.251	9.257	9.263	9.270	9.276	9.282	9.289	9.295
1790	9.302	9.308	9.314	9.321	9.327	9.333	9.340	9.346	9.353	9.359

TABLE A.6. (Continued)

PLAT. +10% RHODIUM vs. PLAT. THERMOCOUPLE
Degrees Fahrenheit   Reference Junction 32° F.

°F	0	1	2	3	4	5	6	7	8	9
					Millivolts					
1800	9.365	9.372	9.378	9.384	9.391	9.397	9.404	9.410	9.416	9.423
1810	9.429	9.436	9.442	9.448	9.455	9.461	9.468	9.474	9.480	9.487
1820	9.493	9.500	9.506	9.512	9.519	9.525	9.532	9.538	9.544	9.551
1830	9.557	9.564	9.570	9.576	9.583	9.589	9.596	9.602	9.609	9.615
1840	9.621	9.628	9.634	9.641	9.647	9.654	9.660	9.666	9.673	9.679
1850	9.686	9.692	9.699	9.705	9.711	9.718	9.724	9.731	9.737	9.744
1860	9.750	9.757	9.763	9.769	9.776	9.782	9.789	9.795	9.802	9.808
1870	9.815	9.821	9.828	9.834	9.840	9.847	9.853	9.860	9.866	9.873
1880	9.879	9.886	9.892	9.899	9.905	9.912	9.918	9.925	9.931	9.937
1890	9.944	9.950	9.957	9.963	9.970	9.976	9.983	9.989	9.996	10.002
1900	10.009	10.015	10.022	10.028	10.035	10.041	10.048	10.054	10.061	10.067
1910	10.074	10.080	10.087	10.093	10.100	10.106	10.113	10.119	10.126	10.132
1920	10.139	10.145	10.152	10.158	10.165	10.171	10.178	10.184	10.191	10.197
1930	10.204	10.210	10.217	10.223	10.230	10.237	10.243	10.250	10.256	10.263
1940	10.269	10.276	10.282	10.289	10.295	10.302	10.308	10.315	10.321	10.328
1950	10.334	10.341	10.348	10.354	10.361	10.367	10.374	10.380	10.387	10.393
1960	10.400	10.406	10.413	10.420	10.426	10.433	10.439	10.446	10.452	10.459
1970	10.465	10.472	10.478	10.485	10.492	10.498	10.505	10.511	10.518	10.524
1980	10.531	10.538	10.544	10.551	10.557	10.564	10.570	10.577	10.583	10.590
1990	10.597	10.603	10.610	10.616	10.623	10.629	10.636	10.643	10.649	10.656
2000	10.662	10.669	10.675	10.682	10.689	10.695	10.702	10.708	10.715	10.722
2010	10.728	10.735	10.741	10.748	10.754	10.761	10.768	10.774	10.781	10.787
2020	10.794	10.801	10.807	10.814	10.820	10.827	10.834	10.840	10.847	10.853
2030	10.860	10.866	10.873	10.880	10.886	10.893	10.899	10.906	10.913	10.919
2040	10.926	10.932	10.939	10.946	10.952	10.959	10.966	10.972	10.979	10.985
2050	10.992	10.999	11.005	11.012	11.018	11.025	11.032	11.038	11.045	11.051
2060	11.058	11.065	11.071	11.078	11.085	11.091	11.098	11.104	11.111	11.118
2070	11.124	11.131	11.137	11.144	11.151	11.157	11.164	11.171	11.177	11.184
2080	11.190	11.197	11.204	11.210	11.217	11.224	11.230	11.237	11.243	11.250
2090	11.257	11.263	11.270	11.277	11.283	11.290	11.296	11.303	11.310	11.316
2100	11.323	11.330	11.336	11.343	11.350	11.356	11.363	11.369	11.376	11.383
2110	11.389	11.396	11.403	11.409	11.416	11.423	11.429	11.436	11.443	11.449
2120	11.456	11.462	11.469	11.476	11.482	11.489	11.496	11.502	11.509	11.516
2130	11.522	11.529	11.536	11.542	11.549	11.556	11.562	11.569	11.575	11.582
2140	11.589	11.595	11.602	11.609	11.615	11.622	11.629	11.635	11.642	11.649
2150	11.655	11.662	11.669	11.675	11.682	11.689	11.695	11.702	11.709	11.715
2160	11.722	11.729	11.735	11.742	11.749	11.755	11.762	11.769	11.775	11.782
2170	11.789	11.795	11.802	11.809	11.815	11.822	11.829	11.835	11.842	11.848
2180	11.855	11.862	11.868	11.875	11.882	11.888	11.895	11.902	11.908	11.915
2190	11.922	11.928	11.935	11.942	11.949	11.955	11.962	11.969	11.975	11.982
2200	11.989	11.995	12.002	12.009	12.015	12.022	12.029	12.035	12.042	12.049
2210	12.055	12.062	12.069	12.075	12.082	12.089	12.095	12.102	12.109	12.115
2220	12.122	12.129	12.135	12.142	12.149	12.155	12.162	12.169	12.175	12.182
2230	12.189	12.196	12.202	12.209	12.216	12.222	12.229	12.236	12.242	12.249
2240	12.256	12.262	12.269	12.276	12.282	12.289	12.296	12.302	12.309	12.316
2250	12.322	12.329	12.336	12.342	12.349	12.356	12.363	12.369	12.376	12.383
2260	12.389	12.396	12.403	12.409	12.416	12.423	12.429	12.436	12.443	12.449
2270	12.456	12.463	12.470	12.476	12.483	12.490	12.496	12.503	12.510	12.516
2280	12.523	12.530	12.536	12.543	12.550	12.556	12.563	12.570	12.577	12.583
2290	12.590	12.597	12.603	12.610	12.617	12.623	12.630	12.637	12.643	12.650
2300	12.657	12.663	12.670	12.677	12.684	12.690	12.697	12.704	12.710	12.717
2310	12.724	12.730	12.737	12.744	12.750	12.757	12.764	12.770	12.777	12.784
2320	12.790	12.797	12.804	12.810	12.817	12.824	12.830	12.837	12.844	12.851
2330	12.857	12.864	12.871	12.877	12.884	12.891	12.897	12.904	12.911	12.917
2340	12.924	12.931	12.937	12.944	12.951	12.957	12.964	12.971	12.977	12.984
2350	12.991	12.997	13.004	13.011	13.018	13.024	13.031	13.038	13.044	13.051
2360	13.058	13.064	13.071	13.078	13.084	13.091	13.098	13.104	13.111	13.118
2370	13.124	13.131	13.138	13.144	13.151	13.158	13.164	13.171	13.178	13.184
2380	13.191	13.198	13.204	13.211	13.218	13.224	13.231	13.238	13.244	13.251
2390	13.258	13.265	13.271	13.278	13.285	13.291	13.298	13.305	13.311	13.318

TABLE A.6. (Continued)

PLAT. +10% RHODIUM vs. PLAT. THERMOCOUPLE
Degrees Fahrenheit   Reference Junction 32° F.

°F	0	1	2	3	4	5	6	7	8	9
					Millivolts					
2400	13.325	13.331	13.338	13.345	13.351	13.358	13.365	13.371	13.378	13.385
2410	13.391	13.398	13.405	13.411	13.418	13.425	13.431	13.438	13.445	13.451
2420	13.458	13.465	13.471	13.478	13.485	13.491	13.498	13.505	13.511	13.518
2430	13.525	13.531	13.538	13.545	13.551	13.558	13.565	13.571	13.578	13.585
2440	13.591	13.598	13.605	13.611	13.618	13.625	13.631	13.638	13.645	13.651
2450	13.658	13.665	13.671	13.678	13.685	13.691	13.698	13.705	13.711	13.718
2460	13.725	13.731	13.738	13.745	13.751	13.758	13.765	13.771	13.778	13.785
2470	13.791	13.798	13.805	13.811	13.818	13.825	13.831	13.838	13.845	13.851
2480	13.858	13.865	13.871	13.878	13.885	13.891	13.898	13.905	13.911	13.918
2490	13.924	13.931	13.938	13.944	13.951	13.958	13.964	13.971	13.978	13.984
2500	13.991	13.998	14.004	14.011	14.018	14.024	14.031	14.038	14.044	14.051
2510	14.058	14.064	14.071	14.078	14.084	14.091	14.098	14.104	14.111	14.118
2520	14.124	14.131	14.137	14.144	14.151	14.157	14.164	14.171	14.177	14.184
2530	14.191	14.197	14.204	14.211	14.217	14.224	14.231	14.237	14.244	14.251
2540	14.257	14.264	14.271	14.277	14.284	14.290	14.297	14.304	14.310	14.317
2550	14.324	14.330	14.337	14.344	14.350	14.357	14.364	14.370	14.377	14.384
2560	14.390	14.397	14.403	14.410	14.417	14.423	14.430	14.437	14.443	14.450
2570	14.457	14.463	14.470	14.477	14.483	14.490	14.497	14.503	14.510	14.516
2580	14.523	14.530	14.536	14.543	14.550	14.556	14.563	14.570	14.576	14.583
2590	14.589	14.596	14.603	14.609	14.616	14.623	14.629	14.636	14.643	14.649
2600	14.656	14.663	14.669	14.676	14.682	14.689	14.696	14.702	14.709	14.716
2610	14.722	14.729	14.736	14.742	14.749	14.755	14.762	14.769	14.775	14.782
2620	14.789	14.795	14.802	14.809	14.815	14.822	14.828	14.835	14.842	14.848
2630	14.855	14.862	14.868	14.875	14.881	14.888	14.895	14.901	14.908	14.915
2640	14.921	14.928	14.935	14.941	14.948	14.954	14.961	14.968	14.974	14.981
2650	14.988	14.994	15.001	15.007	15.014	15.021	15.027	15.034	15.041	15.047
2660	15.054	15.060	15.067	15.074	15.080	15.087	15.094	15.100	15.107	15.113
2670	15.120	15.127	15.133	15.140	15.147	15.153	15.160	15.166	15.173	15.180
2680	15.186	15.193	15.200	15.206	15.213	15.219	15.226	15.233	15.239	15.246
2690	15.253	15.259	15.266	15.272	15.279	15.286	15.292	15.299	15.305	15.312
2700	15.319	15.325	15.332	15.339	15.345	15.352	15.358	15.365	15.372	15.379
2710	15.385	15.391	15.398	15.405	15.411	15.418	15.425	15.431	15.438	15.444
2720	15.451	15.458	15.464	15.471	15.477	15.484	15.491	15.497	15.504	15.510
2730	15.517	15.524	15.530	15.537	15.544	15.550	15.557	15.563	15.570	15.577
2740	15.583	15.590	15.596	15.603	15.610	15.616	15.623	15.629	15.636	15.643
2750	15.649	15.656	15.662	15.669	15.676	15.682	15.689	15.695	15.702	15.709
2760	15.715	15.722	15.728	15.735	15.742	15.748	15.755	15.761	15.768	15.775
2770	15.781	15.788	15.794	15.801	15.808	15.814	15.821	15.827	15.834	15.841
2780	15.847	15.854	15.860	15.867	15.874	15.880	15.887	15.893	15.900	15.907
2790	15.913	15.920	15.926	15.933	15.940	15.946	15.953	15.959	15.966	15.973
2800	15.979	15.986	15.992	15.999	16.006	16.012	16.019	16.025	16.032	16.038
2810	16.045	16.052	16.058	16.065	16.071	16.078	16.085	16.091	16.098	16.104
2820	16.111	16.117	16.124	16.131	16.137	16.144	16.150	16.157	16.164	16.170
2830	16.177	16.183	16.190	16.196	16.203	16.210	16.216	16.223	16.229	16.236
2840	16.243	16.249	16.256	16.262	16.269	16.275	16.282	16.289	16.295	16.302
2850	16.308	16.315	16.322	16.328	16.335	16.341	16.348	16.354	16.361	16.368
2860	16.374	16.381	16.387	16.394	16.400	16.407	16.414	16.420	16.427	16.433
2870	16.440	16.446	16.453	16.460	16.466	16.473	16.479	16.486	16.492	16.499
2880	16.506	16.512	16.519	16.525	16.532	16.538	16.545	16.552	16.558	16.565
2890	16.571	16.578	16.584	16.591	16.597	16.604	16.611	16.617	16.624	16.630
2900	16.637	16.643	16.650	16.657	16.663	16.670	16.676	16.683	16.689	16.696
2910	16.702	16.709	16.716	16.722	16.729	16.735	16.742	16.748	16.755	16.761
2920	16.768	16.775	16.781	16.788	16.794	16.801	16.807	16.814	16.820	16.827
2930	16.834	16.840	16.847	16.853	16.860	16.866	16.873	16.879	16.886	16.893
2940	16.899	16.906	16.912	16.919	16.925	16.932	16.938	16.945	16.952	16.958
2950	16.965	16.971	16.978	16.984	16.991	16.997	17.004	17.010	17.017	17.023
2960	17.030	17.037	17.043	17.050	17.056	17.063	17.069	17.076	17.082	17.089
2970	17.095	17.102	17.109	17.115	17.122	17.128	17.135	17.141	17.148	17.154
2980	17.161	17.167	17.174	17.180	17.187	17.194	17.200	17.207	17.213	17.220
2990	17.226	17.233	17.239	17.246	17.252	17.259	17.265	17.272	17.278	17.285

TABLE A.6. (Continued)

PLAT. +10% RHODIUM vs. PLAT. THERMOCOUPLE
Degrees Fahrenheit   Reference Junction 32° F.

°F	0	1	2	3	4	5	6	7	8	9
					Millivolts					
3000	17.292	17.298	17.305	17.311	17.318	17.324	17.331	17.337	17.344	17.350
3010	17.357	17.363	17.370	17.376	17.383	17.389	17.396	17.402	17.409	17.416
3020	17.422	17.429	17.435	17.442	17.448	17.455	17.461	17.468	17.474	17.481
3030	17.487	17.494	17.500	17.507	17.513	17.520	17.526	17.533	17.539	17.546
3040	17.552	17.559	17.565	17.572	17.578	17.585	17.592	17.598	17.605	17.611
3050	17.618	17.624	17.631	17.637	17.644	17.650	17.657	17.663	17.670	17.676
3060	17.683	17.689	17.696	17.702	17.709	17.715	17.722	17.728	17.735	17.741
3070	17.748	17.754	17.761	17.767	17.774	17.780	17.787	17.793	17.800	17.806
3080	17.813	17.819	17.826	17.832	17.839	17.845	17.852	17.858	17.865	17.871
3090	17.878	17.884	17.891	17.897	17.904	17.910	17.917	17.923	17.930	17.936
3100	17.943	17.949	17.956	17.962	17.669	17.975	17.982	17.988	17.995	18.001
3110	18.008	18.014	18.021	18.027	18.034	18.040	18.047	18.053	18.060	18.066
3120	18.073	18.079	18.086	18.092	18.098	18.105	18.111	18.118	18.124	18.131
3130	18.137	18.144	18.150	18.157	18.163	18.170	18.176	18.183	18.189	18.196
3140	18.202	18.209	18.215	18.222	18.228	18.235	18.241	18.248	18.254	18.260
3150	18.267	18.273	18.280	18.286	18.293	18.299	18.306	18.312	18.319	18.325
3160	18.332	18.338	18.345	18.351	18.358	18.364	18.371	18.377	18.383	18.390
3170	18.396	18.403	18.409	18.416	18.422	18.429	18.435	18.442	18.448	18.455
3180	18.461	18.468	18.474	18.480	18.487	18.493	18.500	18.506	18.513	18.519
3190	18.526	18.532	18.539	18.545	18.551	18.558	18.564	18.571	18.577	18.584
3200	18.590	18.597	18.603	18.610	18.616	18.622	18.629	18.635	18.642	18.648
3210	18.655	18.661	18.668	18.674	18.681	18.687				

Source: National Bureau of Standards Circular 561, 1955. (Reprinted by permission of Leeds & Northrup Company.)

TABLE A.7. **Output of type R thermocouple**

PLAT. +13% RHODIUM vs. PLAT. THERMOCOUPLE
Degrees Fahrenheit   Reference Junction 32° F.

°F	0	1	2	3	4	5	6	7	8	9
					Millivolts					
0	-0.092	-0.089	-0.086	-0.084	-0.081	-0.078	-0.075	-0.072	-0.070	-0.067
10	-.064	-.061	-.058	-.055	-.052	-.050	-.047	-.044	-.041	-.038
20	-.035	-.032	-.029	-.026	-.023	-.021	-.018	-.015	-.012	-.009
30	-.006	-.003	.000	.003	.006	.009	.012	.015	.018	.021
40	.024	.027	.030	.033	.036	.039	.042	.045	.048	.052
50	.055	.058	.061	.064	.068	.071	.074	.077	.080	.083
60	.086	.090	.093	.096	.099	.103	.106	.109	.112	.116
70	.119	.122	.126	.129	.132	.135	.139	.142	.145	.149
80	.152	.155	.159	.162	.165	.169	.172	.175	.179	.182
90	.186	.189	.192	.196	.199	.203	.206	.210	.213	.217
100	.220	.224	.227	.230	.234	.237	.241	.244	.248	.251
110	.255	.258	.262	.265	.269	.272	.276	.280	.284	.287
120	.291	.294	.298	.301	.305	.308	.312	.316	.319	.323
130	.327	.330	.334	.337	.341	.345	.349	.352	.356	.359
140	.363	.367	.370	.374	.378	.381	.385	.389	.393	.397
150	.400	.404	.408	.411	.415	.419	.423	.427	.431	.435
160	.438	.442	.446	.450	.453	.457	.461	.465	.469	.473
170	.476	.480	.484	.488	.492	.496	.500	.504	.508	.512
180	.516	.520	.524	.528	.532	.536	.540	.544	.548	.552
190	.556	.560	.564	.568	.572	.576	.580	.584	.588	.592
200	.596	.600	.604	.608	.612	.616	.620	.625	.629	.633
210	.637	.641	.645	.649	.653	.657	.662	.666	.670	.674
220	.678	.683	.687	.691	.695	.700	.704	.708	.712	.716
230	.721	.725	.729	.734	.738	.742	.746	.750	.755	.759
240	.763	.767	.772	.776	.780	.785	.789	.793	.798	.802
250	.807	.811	.815	.820	.824	.828	.833	.837	.842	.846
260	.850	.855	.859	.863	.868	.872	.877	.881	.886	.890
270	.894	.899	.904	.908	.912	.917	.921	.926	.930	.935
280	.939	.944	.949	.953	.957	.962	.966	.971	.975	.980
290	.984	.989	.993	.998	1.002	1.007	1.011	1.016	1.020	1.025
300	1.030	1.034	1.039	1.043	1.048	1.052	1.057	1.061	1.066	1.071
310	1.075	1.080	1.084	1.089	1.094	1.098	1.103	1.107	1.112	1.117
320	1.121	1.126	1.130	1.135	1.140	1.144	1.149	1.153	1.158	1.163
330	1.167	1.172	1.176	1.181	1.186	1.191	1.195	1.200	1.205	1.210
340	1.214	1.219	1.223	1.228	1.233	1.238	1.242	1.247	1.252	1.257
350	1.261	1.266	1.271	1.276	1.280	1.285	1.290	1.295	1.300	1.304
360	1.309	1.314	1.319	1.323	1.328	1.333	1.338	1.343	1.348	1.352
370	1.357	1.362	1.367	1.372	1.377	1.381	1.386	1.391	1.396	1.401
380	1.406	1.410	1.415	1.420	1.425	1.430	1.435	1.440	1.445	1.450
390	1.455	1.460	1.465	1.470	1.475	1.480	1.484	1.489	1.494	1.499
400	1.504	1.509	1.514	1.519	1.524	1.529	1.533	1.438	1.543	1.548
410	1.553	1.558	1.563	1.568	1.573	1.578	1.583	1.588	1.593	1.598
420	1.603	1.608	1.613	1.618	1.623	1.628	1.633	1.638	1.643	1.648
430	1.653	1.658	1.663	1.668	1.673	1.678	1.683	1.688	1.693	1.698
440	1.703	1.708	1.713	1.719	1.724	1.729	1.734	1.739	1.744	1.749
450	1.754	1.759	1.764	1.769	1.774	1.779	1.785	1.790	1.795	1.800
460	1.805	1.811	1.816	1.821	1.826	1.831	1.836	1.841	1.846	1.851
470	1.856	1.862	1.867	1.872	1.877	1.882	1.887	1.892	1.898	1.903
480	1.908	1.913	1.918	1.924	1.929	1.934	1.939	1.944	1.950	1.955
490	1.960	1.965	1.970	1.976	1.981	1.986	1.991	1.996	2.002	2.007
500	2.012	2.017	2.023	2.028	2.033	2.038	2.044	2.049	2.054	2.059
510	2.065	2.070	2.075	2.081	2.086	2.091	2.096	2.101	2.107	2.112
520	2.117	2.123	2.128	2.133	2.139	2.144	2.149	2.154	2.160	2.165
530	2.170	2.176	2.181	2.186	2.192	2.197	2.202	2.207	2.213	2.218
540	2.223	2.229	2.234	2.239	2.245	2.250	2.255	2.261	2.266	2.271
550	2.277	2.282	2.287	2.293	2.298	2.303	2.308	2.314	2.319	2.325
560	2.330	2.335	2.341	2.346	2.352	2.357	2.363	2.368	2.373	2.379
570	2.384	2.389	2.395	2.401	2.406	2.412	2.417	2.423	2.428	2.433
580	2.438	2.444	2.449	2.455	2.460	2.466	2.471	2.477	2.482	2.487
590	2.493	2.498	2.504	2.509	2.515	2.520	2.526	2.531	2.537	2.542

TABLE A.7. (Continued)

PLAT. +13% RHODIUM vs. PLAT. THERMOCOUPLE
Degrees Fahrenheit   Reference Junction 32° F.

°F	0	1	2	3	4	5	6	7	8	9
					Millivolts					
600	2.547	2.553	2.558	2.564	2.569	2.575	2.580	2.586	2.591	2.597
610	2.602	2.608	2.613	2.619	2.624	2.630	2.635	2.641	2.646	2.652
620	2.657	2.663	2.668	2.674	2.679	2.685	2.690	2.696	2.701	2.707
630	2.712	2.718	2.723	2.729	2.734	2.740	2.746	2.751	2.757	2.762
640	2.768	2.773	2.779	2.784	2.790	2.796	2.801	2.807	2.812	2.818
650	2.823	2.829	2.834	2.840	2.846	2.851	2.857	2.862	2.868	2.873
660	2.879	2.884	2.890	2.896	2.901	2.907	2.912	2.918	2.923	2.929
670	2.935	2.940	2.946	2.952	2.957	2.963	2.968	2.974	2.979	2.985
680	2.991	2.997	3.002	3.008	3.013	3.019	3.024	3.030	3.036	3.041
690	3.047	3.053	3.058	3.064	3.069	3.075	3.081	3.087	3.092	3.098
700	3.103	3.109	3.115	3.120	3.126	3.132	3.137	3.143	3.148	3.154
710	3.160	3.166	3.171	3.177	3.182	3.188	3.194	3.199	3.205	3.211
720	3.217	3.222	3.228	3.234	3.239	3.245	3.251	3.256	3.262	3.268
730	3.273	3.279	3.285	3.291	3.296	3.302	3.308	3.313	3.319	3.325
740	3.330	3.336	3.342	3.348	3.353	3.359	3.365	3.370	3.376	3.382
750	3.387	3.393	3.399	3.405	3.411	3.416	3.422	3.428	3.433	3.439
760	3.445	3.451	3.456	3.462	3.468	3.473	3.479	3.485	3.491	3.497
770	3.502	3.508	3.514	3.519	3.525	3.531	3.537	3.543	3.549	3.554
780	3.560	3.566	3.572	3.577	3.583	3.589	3.595	3.601	3.607	3.612
790	3.618	3.624	3.630	3.635	3.641	3.647	3.653	3.659	3.665	3.671
800	3.677	3.682	3.688	3.694	3.700	3.706	3.712	3.718	3.723	3.729
810	3.735	3.741	3.746	3.752	3.758	3.764	3.770	3.776	3.782	3.788
820	3.794	3.799	3.805	3.811	3.817	3.823	3.829	3.835	3.841	3.846
830	3.852	3.858	3.864	3.870	3.876	3.882	3.888	3.894	3.899	3.905
840	3.911	3.917	3.923	3.929	3.935	3.941	3.946	3.952	3.958	3.964
850	3.970	3.976	3.982	3.988	3.994	3.999	4.005	4.011	4.017	4.023
860	4.029	4.035	4.041	4.047	4.052	4.058	4.064	4.070	4.075	4.081
870	4.087	4.093	4.099	4.105	4.111	4.116	4.122	4.128	4.134	4.140
880	4.146	4.152	4.158	4.164	4.169	4.175	4.181	4.187	4.193	4.199
890	4.205	4.211	4.217	4.223	4.229	4.235	4.241	4.246	4.252	4.258
900	4.264	4.270	4.276	4.282	4.288	4.294	4.300	4.306	4.312	4.318
910	4.324	4.330	4.336	4.342	4.348	4.354	4.360	4.366	4.372	4.378
920	4.384	4.389	4.395	4.401	4.407	4.413	4.419	4.425	4.431	4.437
930	4.443	4.449	4.455	4.461	4.467	4.473	4.479	4.485	4.491	4.497
940	4.503	4.509	4.515	4.521	4.527	4.533	4.539	4.545	4.551	4.557
950	4.563	4.569	4.575	4.581	4.587	4.593	4.599	4.605	4.612	4.618
960	4.624	4.630	4.636	4.642	4.648	4.654	4.660	4.666	4.672	4.679
970	4.685	4.691	4.697	4.703	4.709	4.715	4.721	4.727	4.733	4.740
980	4.746	4.752	4.758	4.764	4.770	4.776	4.782	4.788	4.794	4.801
990	4.807	4.813	4.819	4.825	4.831	4.837	4.844	4.850	4.856	4.862
1000	4.868	4.874	4.881	4.887	4.893	4.899	4.905	4.911	4.917	4.924
1010	4.930	4.936	4.942	4.948	4.954	4.960	4.966	4.972	4.979	4.985
1020	4.991	4.998	5.004	5.010	5.016	5.022	5.028	5.034	5.041	5.047
1030	5.053	5.059	5.066	5.072	5.078	5.084	5.090	5.096	5.102	5.109
1040	5.115	5.121	5.127	5.133	5.139	5.146	5.152	5.158	5.164	5.170
1050	5.176	5.182	5.189	5.195	5.201	5.208	5.214	5.220	5.226	5.232
1060	5.238	5.244	5.251	5.257	5.263	5.270	5.276	5.282	5.288	5.294
1070	5.301	5.307	5.313	5.319	5.326	5.332	5.338	5.344	5.351	5.357
1080	5.363	5.369	5.376	5.382	5.388	5.394	5.401	5.407	5.413	5.419
1090	5.426	5.432	5.438	5.444	5.450	5.457	5.463	5.469	5.476	5.482
1100	5.488	5.494	5.501	5.507	5.513	5.519	5.526	5.532	5.538	5.544
1110	5.551	5.557	5.563	5.570	5.576	5.582	5.589	5.595	5.601	5.607
1120	5.614	5.620	5.626	5.633	5.639	5.645	5.652	5.658	5.664	5.671
1130	5.677	5.684	5.690	5.696	5.703	5.709	5.716	5.722	5.728	5.734
1140	5.741	5.747	5.753	5.760	5.766	5.773	5.779	5.786	5.792	5.798
1150	5.805	5.811	5.817	5.824	5.830	5.837	5.843	5.849	5.856	5.862
1160	5.869	5.875	5.881	5.888	5.894	5.901	5.907	5.913	5.920	5.926
1170	5.933	5.939	5.945	5.952	5.958	5.964	5.971	5.977	5.983	5.990
1180	5.996	6.003	6.009	6.015	6.022	6.028	6.035	6.041	6.047	6.054
1190	6.060	6.067	6.073	6.079	6.086	6.092	6.099	6.105	6.111	6.118

TABLE A.7. (Continued)

PLAT. +13% RHODIUM vs. PLAT. THERMOCOUPLE
Degrees Fahrenheit   Reference Junction 32° F.

°F	0	1	2	3	4	5	6	7	8	9
					Millivolts					
1200	6.125	6.131	6.137	6.143	6.150	6.156	6.163	6.169	6.175	6.182
1210	6.188	6.195	6.201	6.207	6.214	6.220	6.227	6.233	6.239	6.246
1220	6.252	6.259	6.265	6.272	6.278	6.285	6.291	6.298	6.304	6.310
1230	6.317	6.323	6.329	6.336	6.342	6.349	6.355	6.362	6.368	6.375
1240	6.381	6.388	6.394	6.401	6.407	6.414	6.420	6.427	6.433	6.440
1250	6.446	6.453	6.459	6.466	6.472	6.479	6.485	6.492	6.498	6.505
1260	6.511	6.518	6.524	6.531	6.537	6.544	6.550	6.557	6.563	6.570
1270	6.577	6.583	6.589	6.596	6.602	6.609	6.616	6.622	6.629	6.635
1280	6.642	6.648	6.655	6.661	6.668	6.674	6.681	6.687	6.694	6.701
1290	6.707	6.714	6.720	6.727	6.733	6.740	6.746	6.753	6.759	6.766
1300	6.773	6.779	6.786	6.792	6.799	6.805	6.812	6.818	6.825	6.832
1310	6.838	6.845	6.851	6.858	6.865	6.871	6.877	6.884	6.891	6.898
1320	6.904	6.911	6.917	6.924	6.931	6.937	6.943	6.950	6.957	6.964
1330	6.970	6.977	6.983	6.990	6.997	7.003	7.010	7.017	7.023	7.030
1340	7.037	7.043	7.049	7.056	7.063	7.069	7.076	7.083	7.089	7.096
1350	7.103	7.109	7.116	7.123	7.129	7.136	7.143	7.149	7.155	7.162
1360	7.169	7.175	7.182	7.189	7.195	7.202	7.209	7.215	7.222	7.229
1370	7.235	7.242	7.249	7.255	7.262	7.269	7.275	7.282	7.289	7.295
1380	7.302	7.309	7.315	7.322	7.329	7.336	7.342	7.349	7.356	7.362
1390	7.369	7.376	7.382	7.389	7.396	7.403	7.409	7.416	7.423	7.429
1400	7.436	7.443	7.449	7.456	7.463	7.470	7.477	7.483	7.490	7.497
1410	7.503	7.510	7.517	7.523	7.530	7.537	7.544	7.551	7.557	7.564
1420	7.571	7.578	7.585	7.591	7.598	7.605	7.611	7.618	7.625	7.632
1430	7.639	7.645	7.652	7.659	7.665	7.672	7.679	7.686	7.693	7.699
1440	7.706	7.713	7.720	7.727	7.733	7.740	7.747	7.754	7.761	7.767
1450	7.774	7.781	7.788	7.795	7.801	7.808	7.815	7.822	7.829	7.835
1460	7.842	7.849	7.856	7.863	7.870	7.877	7.884	7.891	7.897	7.904
1470	7.911	7.918	7.924	7.931	7.938	7.945	7.952	7.959	7.965	7.972
1480	7.979	7.986	7.993	7.999	8.006	8.013	8.020	8.027	8.033	8.040
1490	8.047	8.054	8.061	8.068	8.075	8.081	8.089	8.095	8.102	8.109
1500	8.116	8.123	8.129	8.136	8.143	8.150	8.157	8.163	8.170	8.177
1510	8.184	8.191	8.198	8.205	8.212	8.218	8.225	8.232	8.239	8.246
1520	8.253	8.260	8.267	8.274	8.281	8.287	8.294	8.301	8.308	8.315
1530	8.322	8.329	8.336	8.343	8.350	8.356	8.363	8.370	8.377	8.384
1540	8.391	8.398	8.405	8.412	8.419	8.426	8.433	8.439	8.446	8.453
1550	8.460	8.467	8.474	8.481	8.488	8.495	8.502	8.509	8.516	8.523
1560	8.530	8.537	8.544	8.551	8.558	8.565	8.571	8.578	8.585	8.592
1570	8.599	8.606	8.613	8.620	8.627	8.634	8.641	8.648	8.655	8.662
1580	8.669	8.676	8.683	8.690	8.697	8.704	8.711	8.718	8.725	8.732
1590	8.739	8.746	8.753	8.760	8.767	8.774	8.781	8.788	8.795	8.802
1600	8.809	8.816	8.823	8.830	8.837	8.844	8.851	8.858	8.865	8.872
1610	8.879	8.886	8.893	8.900	8.907	8.914	8.921	8.928	8.935	8.942
1620	8.949	8.956	8.963	8.970	8.977	8.984	8.991	8.998	9.005	9.012
1630	9.019	9.026	9.033	9.040	9.047	9.054	9.061	9.068	9.075	9.082
1640	9.090	9.097	9.104	9.111	9.118	9.125	9.132	9.139	9.146	9.153
1650	9.161	9.168	9.175	9.182	9.189	9.196	9.203	9.210	9.218	9.225
1660	9.232	9.239	9.246	9.253	9.260	9.267	9.274	9.281	9.289	9.296
1670	9.303	9.310	9.317	9.324	9.331	9.338	9.345	9.353	9.360	9.367
1680	9.374	9.381	9.388	9.395	9.402	9.409	9.416	9.424	9.431	9.438
1690	9.445	9.452	9.459	9.466	9.474	9.481	9.488	9.495	9.502	9.509
1700	9.516	9.523	9.531	9.538	9.545	9.552	9.559	9.566	9.573	9.580
1710	9.587	9.594	9.602	9.609	9.616	9.623	9.630	9.637	9.644	9.651
1720	9.659	9.666	9.673	9.680	9.687	9.694	9.701	9.709	9.716	9.723
1730	9.730	9.737	9.744	9.751	9.759	9.766	9.773	9.780	9.787	9.794
1740	9.802	9.809	9.816	9.823	9.830	9.838	9.845	9.852	9.859	9.866
1750	9.874	9.881	9.888	9.895	9.902	9.910	9.917	9.924	9.931	9.939
1760	9.946	9.953	9.961	9.968	9.975	9.982	9.990	9.997	10.004	10.012
1770	10.019	10.026	10.034	10.041	10.048	10.056	10.063	10.070	10.077	10.084
1780	10.092	10.099	10.106	10.114	10.121	10.129	10.136	10.143	10.150	10.157
1790	10.164	10.172	10.179	10.186	10.194	10.201	10.208	10.215	10.223	10.230

TABLE A.7. (Continued)

PLAT. +13% RHODIUM vs. PLAT. THERMOCOUPLE
Degrees Fahrenheit   Reference Junction 32° F.

°F	0	1	2	3	4	5	6	7	8	9
					Millivolts					
1800	10.237	10.244	10.251	10.259	10.266	10.274	10.281	10.288	10.296	10.303
1810	10.310	10.318	10.325	10.332	10.339	10.347	10.354	10.361	10.369	10.376
1820	10.383	10.391	10.398	10.405	10.412	10.420	10.427	10.434	10.441	10.449
1830	10.456	10.464	10.471	10.478	10.485	10.493	10.500	10.507	10.514	10.522
1840	10.529	10.537	10.544	10.551	10.559	10.566	10.574	10.581	10.588	10.596
1850	10.603	10.610	10.618	10.625	10.632	10.639	10.647	10.654	10.661	10.669
1860	10.676	10.683	10.691	10.698	10.705	10.712	10.720	10.727	10.735	10.742
1870	10.749	10.757	10.764	10.771	10.779	10.786	10.794	10.801	10.809	10.816
1880	10.823	10.831	10.839	10.846	10.854	10.861	10.869	10.876	10.884	10.891
1890	10.898	10.906	10.914	10.921	10.929	10.936	10.944	10.951	10.959	10.966
1900	10.973	10.981	10.988	10.996	11.003	11.011	11.018	11.026	11.033	11.040
1910	11.048	11.055	11.063	11.070	11.078	11.085	11.093	11.100	11.108	11.115
1920	11.122	11.130	11.138	11.145	11.153	11.160	11.168	11.175	11.183	11.190
1930	11.197	11.205	11.213	11.220	11.228	11.235	11.243	11.250	11.258	11.265
1940	11.273	11.280	11.288	11.295	11.303	11.310	11.318	11.325	11.333	11.340
1950	11.348	11.355	11.363	11.371	11.379	11.385	11.393	11.401	11.408	11.416
1960	11.424	11.431	11.439	11.446	11.454	11.461	11.468	11.476	11.484	11.492
1970	11.499	11.507	11.515	11.522	11.529	11.537	11.544	11.552	11.560	11.568
1980	11.575	11.582	11.590	11.598	11.605	11.613	11.620	11.628	11.636	11.643
1990	11.651	11.658	11.666	11.674	11.681	11.689	11.696	11.704	11.712	11.719
2000	11.726	11.734	11.742	11.749	11.757	11.765	11.772	11.779	11.787	11.795
2010	11.802	11.810	11.817	11.825	11.832	11.840	11.848	11.855	11.863	11.871
2020	11.878	11.885	11.893	11.901	11.908	11.916	11.924	11.931	11.938	11.946
2030	11.954	11.961	11.969	11.976	11.984	11.992	11.999	12.007	12.014	12.022
2040	12.029	12.037	12.045	12.052	12.060	12.068	12.075	12.082	12.090	12.098
2050	12.105	12.113	12.121	12.128	12.136	12.144	12.151	12.159	12.166	12.174
2060	12.182	12.189	12.197	12.205	12.212	12.220	12.227	12.235	12.243	12.250
2070	12.258	12.265	12.273	12.281	12.288	12.296	12.304	12.312	12.319	12.327
2080	12.335	12.342	12.350	12.358	12.365	12.373	12.381	12.388	12.396	12.403
2090	12.411	12.419	12.427	12.434	12.442	12.450	12.458	12.465	12.473	12.480
2100	12.488	12.495	12.503	12.511	12.518	12.526	12.534	12.541	12.549	12.557
2110	12.564	12.572	12.579	12.587	12.595	12.602	12.610	12.618	12.625	12.633
2120	12.641	12.648	12.656	12.664	12.672	12.679	12.687	12.695	12.702	12.710
2130	12.718	12.725	12.733	12.741	12.748	12.756	12.764	12.772	12.779	12.787
2140	12.795	12.802	12.810	12.818	12.825	12.833	12.841	12.848	12.856	12.864
2150	12.871	12.879	12.887	12.894	12.902	12.909	12.917	12.925	12.932	12.940
2160	12.948	12.955	12.963	12.971	12.978	12.986	12.994	13.002	13.009	13.017
2170	13.025	13.032	13.040	13.048	13.055	13.063	13.071	13.078	13.086	13.094
2180	13.102	13.109	13.117	13.125	13.132	13.140	13.148	13.155	13.163	13.170
2190	13.178	13.186	13.193	13.201	13.208	13.216	13.224	13.232	13.239	13.247
2200	13.255	13.263	13.270	13.278	13.285	13.293	13.301	13.309	13.316	13.324
2210	13.332	13.340	13.347	13.355	13.363	13.371	13.378	13.386	13.394	13.402
2220	13.409	13.417	13.425	13.432	13.440	13.448	13.455	13.463	13.471	13.479
2230	13.486	13.494	13.502	13.509	13.517	13.525	13.532	13.540	13.548	13.556
2240	13.564	13.571	13.579	13.587	13.595	13.602	13.610	13.618	13.625	13.633
2250	13.641	13.648	13.656	13.664	13.672	13.679	13.687	13.695	13.702	13.710
2260	13.718	13.725	13.733	13.741	13.749	13.756	13.764	13.772	13.779	13.787
2270	13.795	13.802	13.810	13.818	13.826	13.833	13.841	13.849	13.857	13.865
2280	13.872	13.880	13.888	13.895	13.903	13.911	13.918	13.926	13.934	13.942
2290	13.949	13.957	13.965	13.972	13.980	13.988	13.995	14.003	14.011	14.019
2300	14.027	14.034	14.042	14.050	14.058	14.065	14.073	14.081	14.088	14.096
2310	14.104	14.111	14.119	14.127	14.135	14.142	14.150	14.158	14.165	14.173
2320	14.181	14.188	14.196	14.204	14.212	14.219	14.227	14.235	14.242	14.250
2330	14.258	14.265	14.273	14.281	14.288	14.296	14.304	14.311	14.319	14.327
2340	14.335	14.342	14.350	14.358	14.366	14.374	14.382	14.389	14.397	14.405
2350	14.412	14.420	14.428	14.435	14.443	14.451	14.459	14.467	14.475	14.482
2360	14.490	14.498	14.505	14.513	14.521	14.528	14.536	14.544	14.552	14.560
2370	14.567	14.575	14.583	14.591	14.598	14.606	14.614	14.621	14.629	14.637
2380	14.644	14.652	14.660	14.668	14.675	14.683	14.691	14.698	14.706	14.714
2390	14.721	14.729	14.737	14.745	14.752	14.760	14.768	14.775	14.783	14.791

TABLE A.7. (Continued)

PLAT. +13% RHODIUM vs. PLAT. THERMOCOUPLE
Degrees Fahrenheit   Reference Junction 32° F.

°F	0	1	2	3	4	5	6	7	8	9
					Millivolts					
2400	14.798	14.806	14.814	14.822	14.829	14.837	14.845	14.852	14.860	14.868
2410	14.875	14.883	14.891	14.898	14.906	14.914	14.922	14.929	14.937	14.945
2420	14.952	14.960	14.968	14.975	14.983	14.991	14.999	15.006	15.014	15.022
2430	15.029	15.037	15.045	15.052	15.060	15.068	15.076	15.084	15.091	15.099
2440	15.107	15.115	15.122	15.130	15.138	15.145	15.153	15.161	15.168	15.176
2450	15.184	15.192	15.199	15.207	15.215	15.222	15.230	15.238	15.245	15.253
2460	15.261	15.268	15.276	15.284	15.292	15.299	15.307	15.315	15.322	15.330
2470	15.338	15.345	15.353	15.361	15.369	15.377	15.385	15.392	15.400	15.408
2480	15.415	15.423	15.431	15.438	15.446	15.454	15.462	15.469	15.477	15.484
2490	15.492	15.500	15.508	15.515	15.523	15.531	15.538	15.546	15.553	15.561
2500	15.568	15.576	15.584	15.592	15.599	15.607	15.615	15.623	15.630	15.638
2510	15.645	15.653	15.661	15.668	15.676	15.684	15.692	15.700	15.707	15.715
2520	15.722	15.730	15.738	15.745	15.753	15.761	15.769	15.777	15.785	15.792
2530	15.800	15.808	15.815	15.823	15.831	15.838	15.846	15.854	15.862	15.869
2540	15.877	15.885	15.892	15.900	15.908	15.915	15.923	15.931	15.939	15.946
2550	15.954	15.962	15.969	15.977	15.985	15.992	16.000	16.008	16.015	16.023
2560	16.031	16.039	16.046	16.054	16.062	16.070	16.078	16.085	16.093	16.101
2570	16.108	16.116	16.124	16.132	16.139	16.147	16.155	16.163	16.170	16.178
2580	16.185	16.193	16.201	16.208	16.216	16.224	16.232	16.240	16.247	16.255
2590	16.263	16.271	16.278	16.286	16.294	16.301	16.309	16.317	16.325	16.332
2600	16.340	16.348	16.355	16.363	16.371	16.378	16.386	16.394	16.402	16.409
2610	16.417	16.425	16.432	16.440	16.448	16.455	16.463	16.471	16.478	16.486
2620	16.494	16.502	16.509	16.517	16.524	16.532	16.540	16.548	16.556	16.564
2630	16.571	16.579	16.586	16.594	16.602	16.610	16.618	16.625	16.633	16.641
2640	16.648	16.656	16.663	16.671	16.679	16.687	16.695	16.702	16.710	16.718
2650	16.725	16.733	16.741	16.748	16.756	16.764	16.772	16.780	16.799	16.795
2660	16.802	16.810	16.818	16.826	16.834	16.842	16.849	16.857	16.865	16.872
2670	16.880	16.887	16.895	16.903	16.911	16.918	16.926	16.933	16.941	16.949
2680	16.957	16.964	16.972	16.979	16.987	16.995	17.002	17.010	17.018	17.025
2690	17.033	17.041	17.048	17.056	17.064	17.072	17.079	17.087	17.095	17.102
2700	17.110	17.118	17.125	17.133	17.141	17.148	17.756	17.163	17.171	17.179
2710	17.186	17.194	17.202	17.209	17.217	17.225	17.232	17.240	17.248	17.255
2720	17.263	17.271	17.278	17.286	17.294	17.301	17.309	17.317	17.325	17.332
2730	17.340	17.347	17.355	17.363	17.370	17.378	17.385	17.393	17.401	17.408
2740	17.416	17.424	17.432	17.439	17.447	17.455	17.462	17.470	17.478	17.485
2750	17.493	17.500	17.508	17.516	17.524	17.532	17.539	17.546	17.554	17.562
2760	17.569	17.577	17.585	17.592	17.600	17.608	17.615	17.623	17.631	17.638
2770	17.646	17.654	17.662	17.669	17.677	17.685	17.692	17.700	17.708	17.715
2780	17.723	17.731	17.738	17.746	17.753	17.761	17.768	17.776	17.784	17.792
2790	17.799	17.807	17.814	17.822	17.830	17.837	17.845	17.852	17.860	17.868
2800	17.875	17.882	17.890	17.898	17.906	17.913	17.921	17.928	17.936	17.944
2810	17.951	17.958	17.966	17.974	17.982	17.989	17.997	18.004	18.012	18.020
2820	18.027	18.035	18.043	18.050	18.058	18.065	18.073	18.080	18.088	18.096
2830	18.103	18.111	18.119	18.126	18.134	18.141	18.149	18.156	18.164	18.172
2840	18.179	18.187	18.195	18.202	18.210	18.218	18.225	18.233	18.240	18.248
2850	18.255	18.263	18.271	18.278	18.286	18.294	18.301	18.309	18.316	18.324
2860	18.332	18.339	18.347	18.355	18.362	18.370	18.377	18.385	18.392	18.400
2870	18.408	18.415	18.423	18.431	18.438	18.446	18.453	18.461	18.468	18.476
2880	18.484	18.492	18.499	18.507	18.514	18.522	18.529	18.537	18.545	18.552
2890	18.560	18.568	18.575	18.583	18.590	18.598	18.605	18.613	18.621	18.628
2900	18.636	18.644	18.651	18.659	18.666	18.674	18.681	18.689	18.697	18.705
2910	18.712	18.720	18.727	18.735	18.743	18.750	18.758	18.765	18.773	18.781
2920	18.788	18.796	18.803	18.811	18.819	18.826	18.834	18.842	18.849	18.857
2930	18.864	18.872	18.879	18.887	18.895	18.902	18.910	18.918	18.925	18.932
2940	18.940	18.948	18.955	18.963	18.971	18.978	18.986	18.993	19.001	19.008
2950	19.016	19.024	19.031	19.039	19.046	19.054	19.062	19.069	19.077	19.084
2960	19.092	19.099	19.107	19.115	19.122	19.129	19.137	19.145	19.152	19.160
2970	19.168	19.175	19.182	19.190	19.198	19.205	19.213	19.220	19.228	19.235
2980	19.243	19.250	19.258	19.265	19.273	19.281	19.288	19.295	19.303	19.311
2990	19.318	19.326	19.333	19.341	19.348	19.356	19.364	19.371	19.378	19.386

TABLE A.7. (Continued)

PLAT. +13% RHODIUM vs. PLAT. THERMOCOUPLE
Degrees Fahrenheit   Reference Junction 32° F.

°F	0	1	2	3	4	5	6	7	8	9
					Millivolts					
3000	19.394	19.402	19.409	19.417	19.424	19.432	19.439	19.447	19.454	19.462
3010	19.470	19.477	19.485	19.492	19.500	19.508	19.515	19.523	19.530	19.538
3020	19.545	19.553	19.561	19.568	19.576	19.583	19.591	19.598	19.606	19.614
3030	19.621	19.628	19.636	19.644	19.651	19.659	19.667	19.674	19.682	19.689
3040	19.697	19.704	19.712	19.720	19.727	19.735	19.742	19.750	19.758	19.765
3050	19.773	19.780	19.788	19.795	19.803	19.811	19.818	19.826	19.833	19.841
3060	19.848	19.856	19.864	19.871	19.878	19.886	19.894	19.902	19.909	19.916
3070	19.924	19.932	19.939	19.947	19.954	19.962	19.969	19.977	19.984	19.992
3080	19.999	20.007	20.014	20.022	20.029	20.037	20.044	20.052	20.059	20.067
3090	20.075	20.082	20.090	20.097	20.105	20.112	20.120	20.127	20.135	20.142

Source: National Bureau of Standards Circular 561, 1955. (Reprinted by permission of Leeds & Northrup Company.)

TABLE A.8. Temperature equivalents—Orton Standard Pyrometric Cones

LARGE CONES				CONE	SMALL CONES	
(1)60C○	108F○	150C○	270F○	NUMBER	300C○	540F○
585○C.	1085○F.	600○C.	1112○F.	022	*630○C.	*1165○F.
602	1116	614	1137	021	643°C.	1189°F.
625	1157	635	1175	020	666	1231
668	1234	683	1261	019	723	1333
696	1285	717	1323	018	752	1386
727	1341	747	1377	017	784	1443
764	1407	792	1458	016	825	1517
790	1454	804	1479	015	843	1549
834	1533	838	1540	014	*870	*1596
869	1596	852	1566	013	*880	*1615
866	1591	884	1623	012	*900	*1650
886	1627	894	1641	011	*915	*1680
887	1629	894	1641	010	919	1686
915	1679	923	1693	09	955	1751
945	1733	955	1751	08	983	1801
973	1783	984	1803	07	1008	1846
991	1816	999	1830	06	1023	1873
1031	1888	1046	1915	05	1062	1944
1050	1922	1060	1940	04	1098	2008
1086	1987	1101	2014	03	1131	2068
1101	2014	1120	2048	02	1148	2098
1117	2043	1137	2079	01	1178	2152
1136	2077	1154	2109	1	1179	2154
1142	2088	1162	2124	2	1179	2154
1152	2106	1168	2134	3	1196	2185
1168	2134	1186	2167	4	1209	2208
1177	2151	1196	2185	5	1221	2230
1201	2194	1222	2232	6	1255	2291
1215	2219	1240	2264	7	1264	2307
1236	2257	1263	2305	8	1300	2372
1260	2300	1280	2336	9	1317	2403
1285	2345	1305	2381	10	1330	2426
1294	2361	1315	2399	11	1336	2437
1306	2383	1326	2419	12	1355	2471

NOTES:

1. The temperature equivalents in this table apply only to Orton Standard Pyrometric Cones when heated at the rates indicated.

2. Temperature Equivalents are given in degrees Centigrade (°C) and the corresponding degrees Fahrenheit (°F). Rates of heating shown at the head of each column of temperature equivalents are expressed in Centigrade degrees (C°) and Fahrenheit degrees (F°) per hour. These heating rates were maintained uniformly during the last several hundred degrees of temperature rise in the test. All determinations were made in an air atmosphere.

3. Temperature equivalents were not determined for Small Cones Nos. 022, 014, 013, 012, and 011, and for Large Cones No. 37 to 42, inclusive, in the 1956 calibration program. The temperatures shown in the table for the cones noted and

TABLE A.8. (Continued)

LARGE CONES				CONE	P.C.E. CONES	
60C○	108F○	150C○	270F○	NUMBER	150C○	270F○
1306○C.	2383○F.	1326○C.	2419○F.	12	1337○C.	2439○F.
1321	2410	1346	2455	13	1349	2460
1388	2530	1366	2491	14	1398	2548
1424	2595	1431	2608	15	1430	2606
1455	2651	1473	2683	16	1491	2716
1477	2691	1485	2705	17	1512	2754
1500	2732	1506	2743	18	1522	2772
1520	2768	1528	2782	19	1541	2806
1542	2808	1549	2820	20	1564	2847
1586	2887	1590	2894	23	1605	2921
1589	2892	1605	2921	26	1621	2950
1614	2937	1627	2961	27	1640	2984
1614	2937	1633	2971	28	1646	2995
1624	2955	1645	2993	29	1659	3018
1636	2977	1654	3009	30	1665	3029
1661	3022	1679	3054	31	1683	3061
*1685	*3065	*1700	*3092	31½	1699	3090
1706	3103	1717	3123	32	1717	3123
1718	3124	1730	3146	32½	1724	3135
1732	3150	1741	3166	33	1743	3169
1757	3195	1759	3198	34	1763	3205
1784	3243	1784	3243	35	1785	3245
1798	3268	1796	3265	36	1804	3279
ND	ND	ND	ND	37	1820	3308
ND	ND	ND	ND	38	*1850	*3362
ND	ND	ND	ND	39	*1865	*3389
ND	ND	ND	ND	40	*1885	*3425
ND	ND	ND	ND	41	*1970	*3578
ND	ND	ND	ND	42	*2015	*3659

NOTES (Continued):
marked (*) are approximate. Temperature equivalents for P.C.E. Cones Nos. 38
to 42, inclusive (marked *) were determined at a heating rate of 600 C° per
hour. (See Jour. Amer. Cer. Soc. Vol. 9, 1926). All other temperature equivalents
were determined in 1954 at the National Bureau of Standards (See Jour. Amer.
Cer. Soc. Vol. 39, 1956).
    4. The temperature equivalents are not necessarily those at which cones
will deform under firing conditions different from those under which the calibrating determinations were made.
    5. For reproducible results, care should be taken to insure that cones are set
in a plaque with the bending face at the correct angle of 82° from the horizontal, with the cone tips at the correct height above the top of the plaque.
(Large Cones 2″, Small and P.C.E. Cones 15/16″.)

Source: Reprinted by permission of The Edward Orton Jr. Ceramic Foundation.

Cited and selected references

The technical literature pertaining to subjects covered in this book has greatly expanded since publication of the first edition of *Ceramics: Industrial Processing and Testing* in 1972. Literally hundreds of books and conference proceedings, as well as several new serial publications, have been added. This selected listing includes nearly all of the books listed in the first edition, plus approximately 50 additional books published since 1972. In general, only a small fraction of the large number of available conference proceedings have been included here.

Alper, A. M., ed. High temperature oxides. Pt. 1 – Magnesia, lime, and chrome refractories. Pt. 2 – Oxides of rare earths, titanium, zirconium, hafnium, niobium, and tantalum. Pt. 3 – Magnesia, alumina, beryllia ceramics: Fabrication, characterization, and properties. Pt. 4 – Refractory glasses, glass-ceramics, ceramics. In *Refractory materials: A series of monographs,* vol. 5. New York: Academic Press, 1970.

———. Phase diagrams: Materials science and technology. Pt. 1 – Theory, principles, and techniques of phase diagrams. Pt. 2 – The use of phase diagrams in metal, refractory, ceramic, and cement technology. Pt. 3 – The use of phase diagrams in electronic materials and glass technology. In *Refractory materials: A series of monographs,* vol. 6. New York: Academic Press, 1970.

Andrews, A. I. *Ceramics tests and calculations.* New York: John Wiley & Sons, 1928.

———. *Porcelain enamels,* 2d ed. Champaign, Ill.: Twin City Publ. Co., 1945.

Annual book of ASTM standards. Philadelphia: American Society of Testing and Materials.

ASTM manual on quality control of materials. Philadelphia: American Society of Testing and Materials Special Technical Publication 15–C, 1951.

Benedict, R. P. *Fundamentals of temperature, pressure and flow measurement,* 3d ed. New York: John Wiley & Sons, 1984.

Berard, M. F., and Wilder, D. R. *Fundamentals of phase equilibria in ceramic systems.* Marietta, Ohio: R.A.N. Publishers, 1990.

Bergeron, C. J., and Risbud, S. H. *Introduction to phase equilibria in ceramics.* Columbus, Ohio: American Ceramic Society, 1984.

Blackman, L. C. F., ed. *Modern aspects of graphite technology.* New York: Academic Press, 1970.

Blanchere, J. R., and Pettit, F. S. *High temperature corrosion of ceramics.* Park Ridge, N.J.: Noyes Data Corp., 1989.

Brown, G. G. *Unit operations.* New York: John Wiley & Sons, 1950.

Brownell, W. E. *Structural clay products.* New York: Springer-Verlag, 1976.

Buchanan, R. C., ed. *Ceramic Materials for Electronics.* 2d ed. New York: M. Dekker, 1991.

Budnikov, P. P. *The technology of ceramics and refractories.* Translated from Russian by Scripta Technica. Cambridge, Mass.: MIT Press, 1964.

Ceramic data book. Chicago: Cahners Publ. Co. Published annually.

Chand, R. H.; Costello, K. P.; and Payne, T. M. *Finish machining of technical ceramics: An engineering approach.* Paper read at Advanced Ceramics Conference, Rosemont, Illinois, 1990.

Chandler, M. H. *Ceramics in the modern world: Man's first technology comes of age.* Garden City, N.Y.: Doubleday & Co., 1968.

Cheremisinoff, N. P., ed. *Handbook of ceramics and composites.* New York: M. Dekker, 1990.

Clews, F. H. *Heavy clay technology,* 2d ed. New York: Academic Press, 1969.

Conrad, J. W. *Ceramic formulas: The complete compendium; A Guide to clay, glaze, enamel, glass and their colors.* New York: Macmillan, 1973.

Cottrell, A. H. *Dislocations and plastic flow in crystals.* Oxford, England: Clarendon Press, 1953.

Davidge, R. W. *Mechanical behavior of ceramics.* New York: Cambridge University Press, 1979.

Deri, M. *Ferroelectric ceramics.* London: Maclaren & Sons, 1965.

Dinsdale, A., and Moore, F. *Viscosity and its measurement.* London: Physical Society, 1962.

Dyson, B. F.; Lohr, R. D.; and Morrell, R. *Mechanical testing of engineering ceramics at high temperatures.* New York: Elsevier Applied Science, 1989.

Evans, J. W., and De Jonghe, L. C. *The production of inorganic materials.* New York: Macmillan, 1991.

Ford, R. W. *Ceramics Drying.* New York: Pergamon Press, 1986.

Ford, W. F. *Action of heat on clays.* London: Maclaren & Sons, 1965.

_____. *The effect of heat on ceramics.* London: Maclaren & Sons, 1967.

Frechette, V. D. *Failure analysis of brittle materials.* Westerville, Ohio: American Ceramic Society, 1990.

_____., ed. *Non-crystalline solids.* New York: John Wiley & Sons, 1960.

Fulrath, R. M., and Pask, J. A. eds. *Ceramic microstructures.* New York: John Wiley & Sons, 1968.

Grayson, Martin. *Enclyclopedia of glass, ceramics and cement.* New York: Wiley, 1984.

Green, K. J.; and Hannik, R. H. J.; and Swain, M. V. *Transformation toughening of ceramics.* Boca Raton, Fla.: CRC Press, 1989.

Griffiths, R., and Rodford, C. *Calculations in ceramics.* London: Maclaren & Sons, 1968.

Grim, R. E. *Applied clay mineralogy,* 2d ed. New York: McGraw-Hill Book Co., 1968.

Helgenon, C. I. *Ceramic-to-metal bonding.* Boston: Technical Publishers, 1968.

Hench, L. L., and West, J. M. *Principles of electronic ceramics.* New York: Wiley, 1990.

Henry, Edward C. *Electronic ceramics.* Garden City, N.Y.: Doubleday & Co., 1969.

Herbert, J. M. *Ceramics dielectrics and capacitors.* New York: Gordon and Breach, 1985.

Hlavac, Jan. *The technology of glass and ceramics.* New York: Elsevier Scientific Publ. Co., 1983.

Hoerner, Thomas A., and Bear, W. Forrest. *Micrometers, calipers, and gages.* St. Paul, Minn.: Hobar Publications, bull. 169, 1969.

Hove, J. E., and Riley, W. C. *Ceramics for advanced technologies.* New York: John Wiley & Sons, 1965a.

_____. *Modern ceramics.* New York: John Wiley & Sons, 1965b.

Hummel, F. A. *Introduction to phase equilibria in ceramic systems.* New York: M. Dekker, 1984.

Industrial graphite engineering handbook. New York: Carbon Products Div., Union Carbide. Published annually.

Insley, H., and Frechette, V. D. *Microscopy of ceramics and cements.* New York: Academic Press, 1955.

Institute of Ceramics. *Health and safety in ceramics: A guide for educational workshops and studios.* New York: Pergamon Press, 1986.

Jackson, George. *Introduction to whitewares.* London: Maclaren & Sons, 1969.

Jaffe, B.; Cook, W. R., Jr.; and Jaffe, H. *Piezoelectric ceramics.* Marietta, Ohio: R.A.N. Publishers, 1990.

Kingery, W. D. *Property measurements at high temperatures.* New York: John Wiley & Sons, 1959.

_____. *Introduction to ceramics.* New York: John Wiley & Sons, 1960.

_____., ed. *Kinetics of high-temperature processes.* New York: John Wiley & Sons, 1960.

_____., ed. *The changing roles of ceramics in society: 26,000 B.C. to the present.* Westerville, Ohio: American Ceramic Society, 1990.

Kingery, W. D.; Bowen, H. K.; and Uhlmann, D. R. *Introduction to ceramics,* 2d ed. New York: Wiley, 1976.

Kirchner, H. P. *Strengthening of ceramics: Treatments, tests, and design applications.* New York: M. Dekker, 1979.

Lambe, T. W. *Soil testing for engineers.* New York: John Wiley & Sons, 1951.

Lawrence, W. G., ed. *Clay-water systems.* Alfred, State Univ. of N.Y.: College of Ceramics, 1965.

Levin, E. M.; Robbins, C. R.; and McMurdie, H. F. *Phase diagrams for ceramists.* Columbus, Ohio: American Ceramic Society. Volumes published at irregular intervals.

Levinson, L. M., ed. *Electronic ceramics: Properties, devices and applications.* New York: M. Dekker, 1988.

Lewis, M. H., ed. *Glasses and glass ceramics.* New York: Chapman and Hall, 1989.

Lewis, W. K.; Radasch, A. H.; and Lewis, H. C. *Industrial stoichiometry.* New York: McGraw-Hill Book Co., 1954.

McAdams, W. H. *Heat transmission,* 3d ed. New York: McGraw-Hill, 1954.

McColm, I. J., and Clark, N. J. *Forming, shaping and working of high-performance ceramics.* New York: Chapman and Hall, 1988.

McColm, I. J., and Hill, L. *Ceramic science for materials technologists.* New York: Chapman and Hall, 1983.

McCreight, L. R.; Rauch, H. W., Sr.; and Sutton, W. H. *Ceramic and graphite fibers and whiskers: A survey of the technology.* New York: Academic Press, 1966.

McGee, T. D. *Principles and methods of temperature measurement.* New York: John Wiley & Sons, 1988.

Mackenzie, J. D., ed. *Modern aspects of the vitreous state,* vol. 1, 2, 3. Washington: Butterworth & Co., 1960.

McMillan, P. W. Glass-ceramics. In J. P. Roberts and P. Popper, eds., *Nonmetallic solids,* vol. 1. New York: Academic Press, 1964.

————. *Glass-ceramics.* New York: Academic Press, 1979.

McNamara, E. P. *Ceramics,* vol. 1, 2, 3. State College, Penn.: 1938.

Maloney, F. J. Terence. *Glass in the modern world.* Garden City, N.Y.: Doubleday & Co., 1968.

Maskall, K. A., and White, D. *Vitreous enamelling: A guide to modern enamelling practice.* New York: Pergamon Press, 1986.

Mechanical behavior of brittle solids. National Bureau of Standards Monograph 59, 1963.

Microstructure of ceramic materials. National Bureau of Standards Misc. Publ. 257, April 6, 1964.

Mitchell, Lane. *Ceramics: Stone age to space age.* New York: McGraw-Hill Book Co., 1963.

Mitchell, V. *Advanced ceramics: Solving problems and cutting costs.* London: Financial Times Business Information, Ltd., 1987.

Modern refractory practice, 4th ed. Pittsburgh: Harbison-Walker Refractories Co., 1961.

Moore, F. *Rheology of ceramic systems.* London: Maclaren & Sons, 1965.

Morey, G. W. *The properties of glass.* New York: Reinhold Publ., 1954.

Morrell, R. *Handbook of properties of technical and engineering ceramics.* London: Her Majesty's Stationery Office, 1985.

Moulson, A. J., and Herbert, J. M. *Electroceramics: Materials, properties, applications.* New York: Chapman and Hall, 1990.

Mysels, Karol J. *Introduction to colloid chemistry,* 2d ed. New York: Interscience Publishers, 1964.

Nelson, G. C. *Ceramics,* 2d ed. New York: Holt, Rinehart, & Winston, 1971.

Newcomb, R., Jr. *Ceramic whitewares.* New York: Pitman Publ., 1947.

Nordyke, J. S., ed. *Lead in the world of ceramics: A source book for scientists, engineers and students.* Columbus, Ohio: American Ceramic Society, 1984.

North American combustion handbook, 3d ed. Cleveland, Ohio: North American Mfg. Co., 1986.

Norton, F. H. *Ceramics for the artist potter.* Reading, Mass.: Addison-Wesley Publ. Co., 1956.

————. *Refractories,* 4th ed. New York: McGraw-Hill Book Co., 1968.

_____. *Fine ceramics.* New York: McGraw-Hill Book Co., 1970.

_____. *Elements of ceramics,* 2d ed. Reading, Mass.: Addison-Wesley Publ. Co., 1974.

O'Bannon, L. S. *Dictionary of ceramic science and engineering.* New York: Plenum Press, 1984.

Parmelee, Cullen W. *Ceramic glazes,* 3d ed. Rev. Cameron G. Harmon. Chicago: Cahners Publ. Co., 1973.

Perkins, W. A., ed. *Ceramic glossary.* Columbus, Ohio: American Ceramic Society, 1984.

Perry, R. H., and Chilton, C. H., eds. *Chemical engineers' handbook,* 5th ed. New York: McGraw-Hill Book Co., 1973.

Persson, R. *Flat glass technology.* London: Butterworth & Co., 1969.

Phillips, C. J. *Glass: The miracle maker.* New York: Pitman Publ., 1941.

Pober, R. L.; Barringer, E. A.; and Bowen, H. K. *Ceramic technology for electronics.* Silver Spring, Md.: International Society for Hybrid Microelectronics, 1984.

Popper, P., ed. *Special ceramics.* New York: Academic Press, 1965.

Randeraat, J. van, and Setterington, R. E., eds. *Piezoelectric ceramics,* 2d ed. London: Mullard, 1974.

Reed, J. S. *Introduction to the principles of ceramic processing.* New York: Wiley, 1988.

Rhodes, D. *Kilns: Design, construction, and operations.* Philadelphia: Chilton Press, 1969.

Richerson, D. W. *Modern ceramic engineering: Properties, processing and use in design.* New York: M. Dekker, 1982.

_____., ed. *Ceramics: Applications in manufacturing.* Dearborn, Mich.: Society of Manufacturing Engineers, 1988.

Ryan, W., and Radford, C. *Whitewares: Testing and quality control.* New York: Pergamon Press, 1987.

Ryshkewitch, E., and Richerson, D. W. *Oxide ceramics: Physical chemistry and technology,* 2d ed. Haskell, N.J.: General Ceramics, 1985.

Saito, Shinroku, ed. *Fine ceramics.* New York: Elsevier, 1988.

Scholes, S. R. *Modern glass practice.* Chicago: Industrial Publications, 1952.

Schwartz, M. M. *Ceramic joining.* Materials Park, Ohio: ASM International, 1990.

Searle, A. B., and Grimshaw, R. W. *The chemistry and physics of clays and other ceramic materials,* 4th ed. New York: John Wiley & Sons, 1971.

Segal, David. *Chemical synthesis of advanced ceramic materials.* New York: Cambridge University Press, 1989.

Singer, F., and Singer, S. S. *Industrial ceramics.* New York: Chemical Publishing Co., Inc., 1964.

Somiya, S. *Advanced technical ceramics.* Tokyo: Academic Press, 1989.

Somiya, S.; Mitomo, M.; and Yoshimura, M. *Silicon nitride.* New York: Elsevier Applied Science, 1990.

Sosman, R. B. *The phases of silica.* New Brunswick, N. J.: Rutgers University Press, 1965.

Stafford, Eugene. *Modern industrial ceramics.* Indianapolis, Ind.: Bobbs-Merrill Educational Publ., 1980.

Stanworth, J. E. *Physical properties of glass.* London: Oxford Univ. Press, 1950.

Storms, Edmund K., ed. The refractory carbides. In *Refractory materials: A series of monographs,* vol. 2. New York: Academic Press, 1967.

Tallan, N. M., ed. *Electrical conductivity in ceramics and glass.* New York: M. Dekker, 1974.

Taylor, H. F. W. *The chemistry of cements,* vol. 1. New York: Academic Press, 1964.

Taylor, J. R., and Bull, R. C. *Ceramic glaze technology.* New York: Pergamon Press, 1986.

U.S. Military Standard 414—*Sampling procedures and tables for inspection by variables for percent defective.* Washington: USGPO, June 1957.

U.S. Military Standard 105A—*Sampling procedures and tables for inspection by attributes.* Washington: USGPO, 18 July 1961.

Van Schoik, E. C. *Ceramic glossary.* Columbus, Ohio: American Ceramic Society, 1963.

Van Vlack, Lawrence H. *Physical ceramics for engineers.* Reading, Mass.: Addison-Wesley Publ. Co., 1964.

Wachtman, J. B., Jr., ed. *Structural ceramics.* Boston: Academic Press, 1989.

Wang, F. F. Y., ed. *Ceramic fabrication processes.* New York: Academic Press, 1976.

Warren, B. E., *X-ray diffraction.* Reading, Mass.: Addison-Wesley Publ. Co., 1969.

Waye, Basil E. *Introduction to technical ceramics.* London: Maclaren & Sons, 1967.

Winchell, N. H., and Winchell, A. N. *Elements of optical mineralogy,* vols. 1, 2, 3. New York: John Wiley & Sons, 1927.

Worrall, W. E. *Clays: Their nature, origin and general properties.* London: Maclaren & Sons, 1968.

_____. *Ceramic raw materials,* 2d ed. New York: Pergamon Press, 1982.

INDEX